Math Mammoth
Grade 6-B Worktext

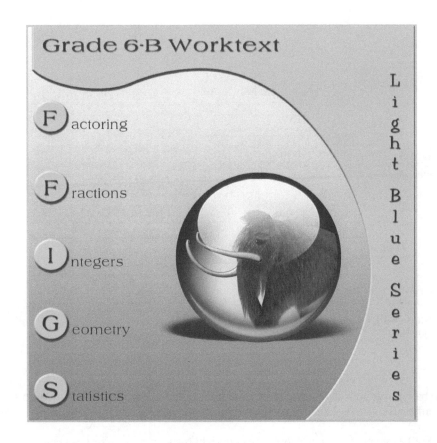

By Maria Miller

Copyright 2022 Taina Maria Miller
ISBN 978-1-954358-07-2

2022 EDITION

All rights reserved. No part of this book may be reproduced or transmitted in any form or by any means, electronic or mechanical, or by any information storage and retrieval system, without permission in writing from the author.

Copying permission: For having purchased this book, the copyright owner grants to the teacher-purchaser a limited permission to reproduce this material for use with his or her students. In other words, the teacher-purchaser MAY make copies of the pages, or an electronic copy of the PDF file, and provide them at no cost to the students he or she is actually teaching, but not to students of other teachers. This permission also extends to the spouse of the purchaser, for the purpose of providing copies for the children in the same family. Sharing the file with anyone else, whether via the Internet or other media, is strictly prohibited.

No permission is granted for resale of the material.

The copyright holder also grants permission to the purchaser to make electronic copies of the material for back-up purposes.

If you have other needs, such as licensing for a school or tutoring center, please contact the author at
https://www.MathMammoth.com/contact.php

Contents

Foreword 5
User Guide 7

Chapter 6: Prime Factorization, GCF and LCM

Introduction 11
The Sieve of Eratosthenes and Prime Factorization 13
Using Factoring When Simplifying Fractions 17
The Greatest Common Factor (GCF) 20
Factoring Sums 23
The Least Common Multiple (LCM) 26
Chapter 6 Mixed Review 30
Chapter 6 Review 32

Chapter 7: Fractions

Introduction 35
Review: Add and Subtract Fractions
and Mixed Numbers 37
Add and Subtract Fractions: More Practice 41
Review: Multiplying Fractions 1 44
Review: Multiplying Fractions 2 47
Dividing Fractions: Reciprocal Numbers 50
Divide Fractions 55
Problem Solving with Fractions 1 59
Problem Solving with Fractions 2 62
Chapter 7 Mixed Review 65
Fractions Review 67

Chapter 8: Integers

Introduction 71
Integers 73
Coordinate Grid 76
Coordinate Grid Practice 80
Addition and Subtraction as Movements 83
Adding Integers: Counters 86

Subtracting a Negative Integer	89
Add and Subtract Roundup	91
Graphing	93
Chapter 8 Mixed Review	97
Integers Review	99

Chapter 9: Geometry

Introduction	103
Quadrilaterals Review	105
Triangles Review	108
Area of Right Triangles	110
Area of Parallelograms	112
Area of Triangles	115
Polygons in the Coordinate Grid	117
Area of Polygons	120
Area of Shapes Not Drawn on Grid	122
Area and Perimeter Problems	124
Nets and Surface Area 1	126
Nets and Surface Area 2	129
Problems to Solve – Surface Area	131
Converting Between Area Units	133
Volume of a Rectangular Prism with Sides of Fractional Length	135
Volume Problems	138
Chapter 9 Mixed Review	140
Geometry Review	143

Chapter 10: Statistics

Introduction	147
Understanding Distributions	149
Mean, Median and Mode	154
Using Mean, Median and Mode	156
Range and Interquartile Range	158
Boxplots	160
Mean Absolute Deviation	163
Making Histograms	167
Summarising Statistical Distributions	170
Stem-and-Leaf-Plots	174
Chapter 10 Mixed Review	176
Statistics Review	179

Foreword

Math Mammoth Grade 6 comprises a complete math curriculum for the sixth grade mathematics studies. The curriculum meets and exceeds the Common Core standards.

In sixth grade, we have quite a few topics to study. Some of them, such as fractions and decimals, students are familiar with, but many others are introduced for the first time (e.g. exponents, ratios, percent, integers).

The main areas of study in Math Mammoth Grade 6 are:

- An introduction to several algebraic concepts, such as exponents, expressions, and equations;
- Rational numbers: fractions, decimals, and percents;
- Ratios, rates, and problem solving using bar models;
- Geometry: area, volume, and surface area;
- Integers and graphing;
- Statistics: summarizing distributions using measures of center and variability.

This book, 6-B, covers number theory topics (chapter 6), fractions (chapter 7), integers (chapter 8), geometry (chapter 9), and statistics (chapter 10). The rest of the topics are covered in the 6-A worktext.

Chapter 6 first reviews prime factorization and then applies those principles to using the greatest common factor to simplify fractions and the least common multiple to find common denominators. Chapter 7 provides a thorough review of the fraction operations from fifth grade, and includes ample practice in solving problems with fractions.

Chapter 8 introduces students to integers. Students plot points in all four quadrants of the coordinate plane, reflect and translate simple figures, and learn to add and subtract with negative numbers. (Multiplication and division of integers will be studied in 7th grade.)

The next chapter, Geometry, focuses on calculating the area of polygons. The final chapter is about statistics. Beginning with the concept of a statistical distribution, students learn about measures of center and measures of variability. They also learn how to make dot plots, histograms, and boxplots, as ways to summarize and analyze distributions.

I heartily recommend that you read the full user guide in the following pages.

I wish you success in teaching math!

Maria Miller, the author

User Guide

Note: You can also find the information that follows online, at https://www.mathmammoth.com/userguides/ .

Basic principles in using Math Mammoth Complete Curriculum

Math Mammoth is mastery-based, which means it concentrates on a few major topics at a time, in order to study them in depth. The two books (parts A and B) are like a "framework", but you still have a lot of liberty in planning your child's studies. You can even use it in a *spiral* manner, if you prefer. Simply have your student study in 2-3 chapters simultaneously. In sixth grade, chapters 1 and 2 should be studied before the other chapters, but you can be flexible with all the other chapters and schedule them earlier or later.

Math Mammoth is not a scripted curriculum. In other words, it is not spelling out in exact detail what the teacher is to do or say. Instead, Math Mammoth gives you, the teacher, various tools for teaching:

- **The two student worktexts** (parts A and B) contain all the lesson material and exercises. They include the explanations of the concepts (the teaching part) in blue boxes. The worktexts also contain some advice for the teacher in the "Introduction" of each chapter.

  The teacher can read the teaching part of each lesson before the lesson, or read and study it together with the student in the lesson, or let the student read and study on his own. If you are a classroom teacher, you can copy the examples from the "blue teaching boxes" to the board and go through them on the board.

- There are hundreds of **videos** matched to the curriculum available at https://www.mathmammoth.com/videos/ . There isn't a video for every lesson, but there are dozens of videos for each grade level. You can simply have the author teach your child or student!

- Don't automatically assign all the exercises. Use your judgment, trying to assign just enough for your student's needs. You can use the skipped exercises later for review. For most students, I recommend to start out by assigning about half of the available exercises. Adjust as necessary.

- For each chapter, there is a **link list to various free online games** and activities. These games can be used to supplement the math lessons, for learning math facts, or just for some fun. Each chapter introduction (in the student worktext) contains a link to the list corresponding to that chapter.

- The student books contain some **mixed review lessons**, and the curriculum also provides you with additional **cumulative review lessons**.

- There is a **chapter test** for each chapter of the curriculum, and a comprehensive end-of-year test.

- The **worksheet maker** allows you to make additional worksheets for most calculation-type topics in the curriculum. This is a single html file. You will need Internet access to be able to use it.

- You can use the free online exercises at https://www.mathmammoth.com/practice/
  This is an expanding section of the site, so check often to see what new topics we are adding to it!

- Some grade levels have **cut-outs** to make fraction manipulatives or geometric solids.

- And of course there are answer keys to everything.

How to get started

Have ready the first lesson from the student worktext. Go over the first teaching part (within the blue boxes) together with your child. Go through a few of the first exercises together, and then assign some problems for your child to do on their own.

Repeat this if the lesson has other blue teaching boxes. Naturally, you can also use the videos at
https://www.mathmammoth.com/videos/

Many children can eventually study the lessons completely on their own — the curriculum becomes self-teaching. However, children definitely vary in how much they need someone to be there to actually teach them.

Pacing the curriculum

The lessons in Math Mammoth complete curriculum are NOT intended to be done in a single teaching session or class. Sometimes you might be able to go through a whole lesson in one day, but more often, the lesson itself might span 3-5 pages and take 2-3 days or classes to complete.

Therefore, it is not possible to say exactly how many pages a student needs to do in one day. This will vary. However, it is helpful to calculate a general guideline as to how many pages per week you should cover in the student worktext in order to go through the curriculum in one school year (or whatever span of time you want to allot to it).

The table below lists how many pages there are for the student to finish in this particular grade level, and gives you a guideline for how many pages per day to finish, assuming a 180-day school year.

Example:

Grade level	Lesson pages	Number of school days	Days for tests and reviews	Days for the student book	Pages to study per day	Pages to study per week
6-A	166	92	10	82	2	10
6-B	157	88	10	79	2	10
Grade 6 total	323	180	20	160	2	10

The table below is for you to fill in. First fill in how many days of school you intend to have. Also allow several days for tests and additional review before the test — at least twice the number of chapters in the curriculum. For example, if the particular grade has 8 chapters, allow at least 16 days for tests & additional review. Then, to get a count of "pages/day", divide the number of pages by the number of available days. Then, multiply this number by 5 to get the approximate page count to cover in a week.

Grade level	Lesson pages	Number of school days	Days for tests and reviews	Days for the student book	Pages to study per day	Pages to study per week
6-A	166					
6-B	157					
Grade 6 total	323					

Now, let's assume you determine that you need to study about 2 pages a day, 10 pages a week in order to get through the curriculum. As you study each lesson, keep in mind that sometimes most of the page might be filled with blue teaching boxes and very few exercises. You might be able to cover 3 pages on such a day. Then some other day you might only assign one page of word problems. Also, you might be able to go through the pages quicker in some chapters, for example when studying graphs, because the large pictures fill the page so that one page does not have many problems.

When you have a page or two filled with lots of similar practice problems ("drill") or large sets of problems, feel free to **only assign 1/2 or 2/3 of those problems**. If your child gets it with less amount of exercises, then that is perfect! If not, you can always assign him/her the rest of the problems some other day. In fact, you could even use these unassigned problems the next week or next month for some additional review.

In general, 1st-2nd graders might spend 25-40 minutes a day on math. Third-fourth graders might spend 30-60 minutes a day. Fifth-sixth graders might spend 45-75 minutes a day. If your child finds math enjoyable, he/she can of course spend more time with it! However, it is not good to drag out the lessons on a regular basis, because that can then affect the child's attitude towards math.

Working space, the usage of additional paper and mental math

The curriculum generally includes working space directly on the page for students to work out the problems. However, feel free to let your students to use extra paper when necessary. They can use it, not only for the "long" algorithms (where you line up numbers to add, subtract, multiply, and divide), but also to draw diagrams and pictures to help organize their thoughts. Some students won't need the additional space (and may resist the thought of extra paper), while some will benefit from it. Use your discretion.

Some exercises don't have any working space, but just an empty line for the answer (e.g. 200 + _____ = 1,000). Typically, I have intended that such exercises to be done using MENTAL MATH.

However, there are some students who struggle with mental math (often this is because of not having studied and used it in the past). As always, the teacher has the final say (not me!) as to how to approach the exercises and how to use the curriculum. We do want to prevent extreme frustration (to the point of tears). The goal is always to provide SOME challenge, but not too much, and to let students experience success enough so that they can continue enjoying learning math.

Students struggling with mental math will probably benefit from studying the basic principles of mental calculations from the earlier levels of Math Mammoth curriculum. To do so, look for lessons that list mental math strategies. They are taught in the chapters about addition, subtraction, place value, multiplication, and division. My article at https://www.mathmammoth.com/lessons/practical_tips_mental_math also gives you a summary of some of those principles.

Using tests

For each chapter, there is a **chapter test**, which can be administered right after studying the chapter. **The tests are optional.** Some families might prefer not to give tests at all. The main reason for the tests is for diagnostic purposes, and for record keeping. These tests are not aligned or matched to any standards.

In the digital version of the curriculum, the tests are provided both as PDF files and as html files. Normally, you would use the PDF files. The html files are included so you can edit them (in a word processor such as Word or LibreOffice), in case you want your student to take the test a second time. Remember to save the edited file under a different file name, or you will lose the original.

The end-of-year test is best administered as a diagnostic or assessment test, which will tell you how well the student remembers and has mastered the mathematics content of the entire grade level.

Using cumulative reviews and the worksheet maker

The student books contain mixed review lessons which review concepts from earlier chapters. The curriculum also comes with additional cumulative review lessons, which are just like the mixed review lessons in the student books, with a mix of problems covering various topics. These are found in their own folder in the digital version, and in the Tests & Cumulative Reviews book in the print version.

The cumulative reviews are optional; use them as needed. They are named indicating which chapters of the main curriculum the problems in the review come from. For example, "Cumulative Review, Chapter 4" includes problems that cover topics from chapters 1-4.

Both the mixed and cumulative reviews allow you to spot areas that the student has not grasped well or has forgotten. When you find such a topic or concept, you have several options:

1. Check if the worksheet maker lets you make worksheets for that topic.
2. Check for any online games and resources in the Introduction part of the particular chapter in which this topic or concept was taught.
3. If you have the digital version, you could simply reprint the lesson from the student worktext, and have the student restudy that.
4. Perhaps you only assigned 1/2 or 2/3 of the exercise sets in the student book at first, and can now use the remaining exercises.
5. Check if our online practice area at https://www.mathmammoth.com/practice/ has something for that topic.
6. Khan Academy has free online exercises, articles, and videos for most any math topic imaginable.

Concerning challenging word problems and puzzles

While this is not absolutely necessary, I heartily recommend supplementing Math Mammoth with challenging word problems and puzzles. You could do that once a month, for example, or more often if the student enjoys it.

The goal of challenging story problems and puzzles is to **develop the student's logical and abstract thinking and mental discipline**. I recommend starting these in fourth grade, at the latest. Then, students are able to read the problems on their own and have developed mathematical knowledge in many different areas. Of course I am not discouraging students from doing such in earlier grades, either.

Math Mammoth curriculum contains lots of word problems, and they are usually multi-step problems. Several of the lessons utilize a bar model for solving problems. Even so, the problems I have created are usually tied to a specific concept or concepts. I feel students can benefit from solving problems and puzzles that require them to think "out of the box" or are just different from the ones I have written.

I recommend you use the free Math Stars problem-solving newsletters as one of the main resources for puzzles and challenging problems:

Math Stars Problem Solving Newsletter (grades 1-8)
https://www.homeschoolmath.net/teaching/math-stars.php

I have also compiled a list of other resources for problem solving practice, which you can access at this link:

https://l.mathmammoth.com/challengingproblems

Another idea: you can find puzzles online by searching for "brain puzzles for kids," "logic puzzles for kids" or "brain teasers for kids."

Frequently asked questions and contacting us

If you have more questions, please first check the FAQ at https://www.mathmammoth.com/faq-lightblue

If the FAQ does not cover your question, you can then contact us using the contact form at the Math Mammoth.com website.

I wish you success in teaching math!

Maria Miller, the author

Chapter 6: Prime Factorization, GCF and LCM
Introduction

The topics of this chapter belong to a branch of mathematics known as *number theory*. Number theory has to do with the study of whole numbers and their special properties. In this chapter, we review prime factorization and study the greatest common factor (GCF) and the least common multiple (LCM).

The main application of factoring and the greatest common factor in arithmetic is in simplifying fractions, so that is why I have included a lesson on that topic. However, it is not absolutely necessary to use the GCF when simplifying fractions, and the lesson emphasizes that fact.

The concepts of factoring and the GCF are important to understand because they will be carried over into algebra, where students will factor polynomials. In this chapter, we lay the groundwork for that by using the GCF to factor simple sums, such as 27 + 45. For example, a sum like 27 + 45 factors into 9(3 + 5).

Similarly, the main use for the least common multiple in arithmetic is in finding the smallest common denominator for adding fractions, and we study that topic in this chapter in connection with the LCM.

Primes are fascinating "creatures," and you can let students read more about them by accessing the Internet resources mentioned below. The really important, but far more advanced, application of prime numbers is in cryptography. Some students might be interested in reading additional material on that subject—please see the list for Internet resources.

Keep in mind that the specific lessons in the chapter can take several days to finish. They are not "daily lessons." Instead, use the general guideline that sixth graders should finish about 2 pages daily or 9-10 pages a week in order to finish the curriculum in about 40 weeks. Also, I recommend not assigning all the exercises by default, but that you use your judgment, and strive to vary the number of assigned exercises according to the student's needs. Please see the user guide at **https://www.mathmammoth.com/userguides/** for more guidance on using and pacing the curriculum.

You can find some free videos for the topics of this chapter at **https://www.mathmammoth.com/videos/** (choose 6th grade).

The Lessons in Chapter 6

	page	span
The Sieve of Eratosthenes and Prime Factorization	13	*4 pages*
Using Factoring When Simplifying Fractions	17	*3 pages*
The Greatest Common Factor (GCF)	20	*3 pages*
Factoring Sums	23	*3 pages*
The Least Common Multiple (LCM)	26	*4 pages*
Chapter 6 Mixed Review	30	*2 pages*
Chapter 6 Review	32	*2 pages*

Helpful Resources on the Internet

We have compiled a list of Internet resources that match the topics in this chapter. This list of links includes web pages that offer:

- **online practice** for concepts;
- online **games**, or occasionally, printable games;
- **animations** and interactive **illustrations** of math concepts;
- **articles** that teach a math concept.

We heartily recommend you take a look at the list. Many of our customers love using these resources to supplement the bookwork. You can use the resources as you see fit for extra practice, to illustrate a concept better, and even just for some fun. Enjoy!

https://l.mathmammoth.com/gr6ch6

The Sieve of Eratosthenes and Prime Factorization

Remember? A number is a **prime** if it has no other factors besides 1 and itself.

For example, 13 is a prime, since the only way to write it as a multiplication is 1 · 13. In other words, 1 and 13 are its only factors.

And, 15 is not a prime, since we can write it as 3 · 5. In other words, 15 has other factors besides 1 and 15, namely 3 and 5.

To find all the prime numbers less than 100 we can use the *sieve of Eratosthenes*.

Here is an online interactive version: https://www.mathmammoth.com/practice/sieve-of-eratosthenes

1. Cross out 1, as it is not considered a prime.
2. Cross out all the even numbers except 2.
3. Cross out all the multiples of 3 except 3.
4. You do not have to check multiples of 4. Why?
5. Cross out all the multiples of 5 except 5.
6. You do not have to check multiples of 6. Why?
7. Cross out all the multiples of 7 except 7.
8. You do not have to check multiples of 8 or 9 or 10.
9. The numbers left are primes.

1̸	2	3	4̸	5	6̸	7	8̸	9	10
11	12	13	14	15	16	17	18	19	20
21	22	23	24	25	26	27	28	29	30
31	32	33	34	35	36	37	38	39	40
41	42	43	44	45	46	47	48	49	50
51	52	53	54	55	56	57	58	59	60
61	62	63	64	65	66	67	68	69	70
71	72	73	74	75	76	77	78	79	80
81	82	83	84	85	86	87	88	89	90
91	92	93	94	95	96	97	98	99	100

List the **primes between 0 and 100** below:

2, 3, 5, 7, _____

Why do you not have to check numbers that are bigger than 10? Let's think about multiples of 11. The following multiples of 11 have already been crossed out: 2 · 11, 3 · 11, 4 · 11, 5 · 11, 6 · 11, 7 · 11, 8 · 11 and 9 · 11. The multiples of 11 that have not been crossed out are 10 · 11 and onward... but they are not on our chart! Similarly, the multiples of 13 that are less than 100 are 2 · 13, 3 · 13, ..., 7 · 13, and all of those have already been crossed out when you crossed out multiples of 2, 3, 5 and 7.

1. You learned this in 4th and 5th grades... find all the factors of the given numbers. Use the checklist to help you keep track of which factors you have tested.

a. 54	**b.** 60
Check 1 2 3 4 5 6 7 8 9 10	Check 1 2 3 4 5 6 7 8 9 10
factors: _____	factors: _____
c. 84	**d.** 97
Check 1 2 3 4 5 6 7 8 9 10	Check 1 2 3 4 5 6 7 8 9 10
factors: _____	factors: _____

For your reference, here are some of the common divisibility tests for whole numbers.	
A number is... **divisible by 2** if it ends in 0, 2, 4, 6, or 8. **divisible by 5** if it ends in 0 or 5. **divisible by 10** if it ends in 0. **divisible by 100** if it ends in "00".	A number is... **divisible by 3** if the sum of its digits is divisible by 3. **divisible by 4** if the number formed from its last two digits is divisible by 4. **divisible by 6** if it is divisible by both 2 and 3. **divisible by 9** if the sum of its digits is divisible by 9.

Use the various divisibility tests when building a factor tree for a composite number.

We start out by noticing that 135 is **divisible by 5**. From long division, we get 135 = 5 · 27. The final factorization is 135 = 3 · 3 · 3 · 5 or $3^3 \cdot 5$.

Adding the digits of 441, we get 9, so it is **divisible by 9**. We divide to get 441 = 9 · 49. The end result is 441 = 3 · 3 · 7 · 7 or $3^2 \cdot 7^2$.

2. Find the prime factorization of these composite numbers. Use a notebook for long divisions. Give each factorization below the factor tree.

a. 124

b. 260

c. 96

124 =

260 =

96 =

d. 90

e. 165

f. 95

90 =

165 =

95 =

3. Mark an "x" if the number is divisible by 2, 3, 4, 5, 6, or 9.

Divisible by	2	3	4	5	6	9
128						
765						

Divisible by	2	3	4	5	6	9
209						
6,042						

4. Find the prime factorization of the numbers. Use a notebook for long divisions. Give each factorization below the factor tree.

Note: in (a), the last two digits of 912 are "12" so it is **divisible by 4**.

a. 912
  / \
 4 · ___

912 =

b. 528

528 =

c. 76

76 =

d. 126

126 =

e. 272

272 =

5. Mia and Alex found the prime factorization of 164 and 168, and were completely surprised that they got the same factorization for both!

Investigate the situation. Is there something fishy going on somewhere?

164
 / \
4 · 42
/ \ / \
2 · 2 · 6 · 7
   / \
  2 · 3

$164 = 2^3 \cdot 3 \cdot 7$

168
 / \
8 · 21
/ \ / \
2 · 4 · 3 · 7
 / \
2 · 2

$168 = 2^3 \cdot 3 \cdot 7$

6. Find all the primes between 100 and 110. How? You need to check, for each number, whether it is divisible by 2, 3, 4, 5, 6, 7, 8, 9, or 10.

7. Find the prime factorization of these composite numbers.

a. 196	b. 380	c. 336
196 =	380 =	336 =
d. 306	e. 116	f. 720
306 =	116 =	720 =
g. 675	h. 990	i. 945
675 =	990 =	945 =

Puzzle Corner Find all the primes between 0 and 200. Use the sieve of Eratosthenes again (you need to make a grid in your notebook).

This time, you need to cross out 1, and then every even number except 2, every multiple of 3 except 3, every multiple of 5 except 5, every multiple of 7 except 7, every multiple of 11 except 11 and every multiple of 13 except 13.

Using Factoring When Simplifying Fractions

You have seen the process of **simplifying fractions** before.

In simplifying fractions, we divide both the numerator and the denominator by the same number. The fraction becomes *simpler*, which means that the numerator and the denominator are now *smaller* numbers than they were before.

Every four slices have been joined together.

$$\frac{12}{20} = \frac{3}{5}$$ (÷ 4 on top and bottom)

However, this does NOT change the actual value of the fraction. It is the "same amount of pie" as it was before. It is just cut differently.

Why does this work?

It is based on finding common factors and on how fraction multiplication works. In our example above, the fraction $\frac{12}{20}$ can be written as $\frac{4 \cdot 3}{4 \cdot 5}$. Then we can **cancel out** those fours: $\frac{\cancel{4} \cdot 3}{\cancel{4} \cdot 5} = \frac{3}{5}$.

The reason this works is because $\frac{4 \cdot 3}{4 \cdot 5}$ is equal to the fraction multiplication $\frac{4}{4} \cdot \frac{3}{5}$. And in that, 4/4 is equal to 1, which means we are only left with 3/5.

Example 1. Often, the simplification is simply written or indicated this way →

Notice that here, the 4's that were cancelled out do *not* get indicated in any way! You only think it: "I divide 12 by 4, and get 3. I divide 20 by 4, and get 5."

$$\frac{\overset{3}{\cancel{12}}}{\underset{5}{\cancel{20}}} = \frac{3}{5}$$

Example 2. Here, 35 and 55 are both divisible by 5. This means we can cancel out those 5's, but notice this is not shown in any way. We simply cross out 35 and 55, think of dividing them by 5, and write the division result above and below.

$$\frac{\overset{7}{\cancel{35}}}{\underset{11}{\cancel{55}}} = \frac{7}{11}$$

1. Simplify the fractions, if possible.

a. $\frac{12}{36}$	b. $\frac{45}{55}$	c. $\frac{15}{23}$	d. $\frac{13}{6}$
e. $\frac{15}{21}$	f. $\frac{19}{15}$	g. $\frac{17}{24}$	h. $\frac{24}{30}$

2. Leah simplified various fractions like you see below. She did not get them right though. Explain to her what she is doing wrong.

$$\frac{24}{84} = \frac{20}{80} = \frac{1}{4} \qquad \frac{27}{60} = \frac{7}{40} \qquad \frac{14}{16} = \frac{10}{12} = \frac{6}{8} = \frac{3}{4}$$

Using factoring when simplifying

Carefully study the example on the right where we simplify the fraction 144/96 to **lowest terms** -- in other words, where the numerator and the denominator have no common factors.

- First we factor (write) 144 as 12 · 12 and 96 as 8 · 12.
- Then we simplify in two steps:
    1. 12 and 8 are both divisible by 4, so they simplify into 3 and 2.
    2. 12 and 12 are divisible by 12, so they simplify into 1 and 1. Essentially, they cancel each other out.
- Lastly we write the improper fraction 3/2 as a mixed number.

$$\frac{144}{96} = \frac{\overset{3}{\cancel{12}} \cdot \overset{1}{\cancel{12}}}{\underset{2}{\cancel{8}} \cdot \underset{1}{\cancel{12}}} = \frac{3}{2} = 1\frac{1}{2}$$

For a comparison, here is another way to write the simplification in several steps, and that you've seen in earlier grades in Math Mammoth:

$$\frac{144}{96} \xrightarrow{\div 12} \frac{12}{8} \xrightarrow{\div 4} \frac{3}{2} = 1\frac{1}{2}$$

Let's study some more examples. (Remember that they don't show the number that you divide by.)

$$\frac{42}{105} = \frac{\overset{1}{\cancel{7}} \cdot \overset{2}{\cancel{6}}}{\underset{5}{\cancel{35}} \cdot \underset{1}{\cancel{3}}} = \frac{2}{5}$$

$$\frac{45}{150} = \frac{\overset{3}{\cancel{9}} \cdot \overset{1}{\cancel{5}}}{\underset{10}{\cancel{30}} \cdot \underset{1}{\cancel{5}}} = \frac{3}{10}$$

3. Simplify. Write the simplified numerator above and the simplified denominator below the old ones.

a. $\frac{14}{16}$	b. $\frac{33}{27}$	c. $\frac{12}{26}$	d. $\frac{9}{33}$	e. $\frac{42}{28}$

4. The numerator and the denominator have already been factored in some problems. Your task is to simplify. Also, give your final answer as a mixed number, if applicable.

a. $\frac{56}{84} = \frac{7 \cdot 8}{21 \cdot 4} =$	b. $\frac{54}{144} = \frac{6 \cdot 9}{12 \cdot 12} =$	c. $\frac{120}{72} = \frac{10 \cdot \square}{\square \cdot 9} =$
d. $\frac{80}{48} = \frac{\square \cdot 8}{\square \cdot 8} =$	e. $\frac{36}{90} = \underline{\qquad} =$	f. $\frac{28}{140} = \underline{\qquad} =$

5. Simplify the fractions. Use your knowledge of divisibility.

a. $\frac{95}{100}$	b. $\frac{66}{82}$	c. $\frac{69}{99}$
d. $\frac{120}{600}$	e. $\frac{38}{52}$	f. $\frac{72}{84}$

Simplify "criss-cross"

These examples are from the previous page. This time the 45 in the numerator has been written as 5 · 9 instead of 9 · 5. We can cancel out the 5 from the numerator with the 5 from the denominator (we simplify criss-cross).

$$\frac{45}{150} = \frac{\overset{1}{\cancel{5}} \cdot \overset{3}{\cancel{9}}}{\underset{10}{\cancel{30}} \cdot \underset{1}{\cancel{5}}} = \frac{3}{10}$$

Also, we can simplify the 9 in the numerator and the 30 in the denominator criss-cross. The other example (simplifying 42/105) is similar.

$$\frac{42}{105} = \frac{\overset{1}{\cancel{7}} \cdot \overset{2}{\cancel{6}}}{\underset{1}{\cancel{3}} \cdot \underset{5}{\cancel{35}}} = \frac{2}{5}$$

This same concept can be applied to make multiplying fractions easier.

6. Simplify. Give your final answer as a mixed number, if applicable.

a. $\dfrac{14}{84} = \dfrac{2 \cdot 7}{21 \cdot 4} =$	b. $\dfrac{54}{150} = \dfrac{9 \cdot }{10 \cdot } =$	c. $\dfrac{138}{36} = \dfrac{2 \cdot }{ \cdot 4} =$
d. $\dfrac{27}{20} \cdot \dfrac{10}{21} =$	e. $\dfrac{75}{90} = \underline{} =$	f. $\dfrac{48}{45} \cdot \dfrac{55}{64} =$

Example 3. Here, the simplification is done in two steps. In the first step, 12 and 2 are divided by 2, leaving 6 and 1. In the second step, 6 and 69 are divided by 3, leaving 2 and 23.

$$\frac{48}{138} = \frac{\overset{6}{\cancel{12}} \cdot 4}{\underset{1}{\cancel{2}} \cdot 69} = \frac{\overset{2}{\cancel{6}} \cdot 4}{1 \cdot \underset{23}{\cancel{69}}} = \frac{8}{23}$$

These two steps can also be done without rewriting the expression. The 6 and 69 are divided by 3 as before. This time we simply did not rewrite the expression in between but just continued on with the numbers 6 and 69 that were already written there.

$$\frac{48}{138} = \frac{\overset{\overset{2}{\cancel{6}}}{\cancel{12}} \cdot 4}{\underset{1}{\cancel{2}} \cdot \underset{23}{\cancel{69}}} = \frac{8}{23}$$

If this looks too confusing, you do not have to write it in such a compact manner. You can rewrite the expression before simplifying it some more.

7. Simplify the fractions to lowest terms, or simplify before you multiply the fractions.

a. $\dfrac{88}{100}$	b. $\dfrac{84}{102}$	c. $\dfrac{85}{105}$
d. $\dfrac{8}{5} \cdot \dfrac{8}{20} =$	e. $\dfrac{72}{120}$	f. $\dfrac{104}{240}$
g. $\dfrac{35}{98}$	h. $\dfrac{5}{7} \cdot \dfrac{17}{15} =$	i. $\dfrac{72}{112}$

The Greatest Common Factor (GCF)

Let's take two whole numbers. We can then list all the <u>factors</u> of each number, and then find the factors that are <u>common</u> in both lists. Lastly, we can choose the <u>greatest</u> or largest among those "common factors." That is the **greatest common factor** of the two numbers. The term itself really tells you what it means!

Example 1. Find the greatest common factor of 18 and 30.

<u>The factors of 18:</u> 1, 2, 3, 6, 9 and 18.
<u>The factors of 30:</u> 1, 2, 3, 5, 6, 10, 15 and 30.

Their <u>common</u> factors are 1, 2, 3 and 6. The <u>greatest</u> common factor is 6.

Here is a **method to find all the factors of a given number**.

Example 2. Find the factors (divisors) of 36.

We check if 36 is divisible by 1, 2, 3, 4 and so on. Each time we find a divisor, we write down *two* factors.

- 36 is divisible by 1. We write $36 = 1 \cdot 36$, and that equation gives us two factors of 36: both the smallest (**1**) and the largest (**36**).
- 36 is also divisible by 2. We write $36 = 2 \cdot 18$, and that equation gives us two more factors of 36: the second smallest (**2**) and the second largest (**18**).
- Next, 36 is divisible by 3. We write $36 = 3 \cdot 12$, and now we have found the third smallest factor (**3**) and the third largest factor (**12**).
- Next, 36 is divisible by 4. We write $36 = 4 \cdot 9$, and we have found the fourth smallest factor (**4**) and the fourth largest factor (**9**).
- Finally, 36 is divisible by 6. We write $36 = 6 \cdot 6$, and we have found the fifth smallest factor (**6**) which is also the fifth largest factor.

We know that we are done because the list of factors from the "small" end (1, 2, 3, 4, 6) has met the list of factors from the "large" end (36, 18, 12, 9, 6).

Therefore, all of the factors of 36 are: 1, 2, 3, 4, 6, 9, 12, 18 and 36.

1. List all of the factors of the given numbers.

a. 48	**b.** 60
c. 42	**d.** 99

2. Find the greatest common factor of the given numbers. Your work above will help!

a. 48 and 60	**b.** 42 and 48	**c.** 42 and 60	**d.** 99 and 60

3. List all of the factors of the given numbers.

a. 44	b. 66
c. 28	d. 56
e. 100	f. 45

4. Find the greatest common factor of the given numbers. Your work above will help!

a. 44 and 66	b. 100 and 28	c. 45 and 100	d. 45 and 66
e. 28 and 44	f. 56 and 28	g. 56 and 100	h. 45 and 28

Example 3. What is the greatest common factor useful for?

It can be used to simplify fractions. For example, let's say you know that the GCF of 66 and 84 is 6. Then, to simplify the fraction 66/84 to lowest terms, you divide both the numerator and the denominator by 6. →

However, it is *not* necessary to use the GCF when simplifying fractions. You can always simplify in several steps. See the example at the right. →

Or, you can *simplify by factoring*, like we did in the previous lesson:

$$\frac{66}{84} = \frac{6 \cdot 11}{7 \cdot 6 \cdot 2} = \frac{11}{14}$$

In fact, these other methods might be quicker than using the GCF.

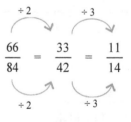

5. Simplify these fractions, if possible. Your work in the previous exercises can help!

a. $\dfrac{48}{66}$ b. $\dfrac{42}{44}$ c. $\dfrac{42}{48}$ d. $\dfrac{99}{60}$

e. $\dfrac{48}{100}$ f. $\dfrac{100}{99}$ g. $\dfrac{56}{28}$ h. $\dfrac{44}{99}$

Using prime factorization to find the greatest common factor (optional)

Another, more efficient way to find the GCF of two or more numbers is to use the prime factorizations of the numbers to find *all* of the common prime factors. The product of those common prime factors forms the GCF.

Example 4. Find the GCF of 48 and 84.

The prime factorizations are: $48 = 2 \cdot 2 \cdot 2 \cdot 2 \cdot 3$ and $84 = 2 \cdot 2 \cdot 3 \cdot 7$.

We see that the common prime factors are 2 and 2 and 3. So, the GCF is $2 \cdot 2 \cdot 3 = 12$.

Example 5. Find the GCF of 75, 105 and 150.

The prime factorizations are: $75 = 3 \cdot 5 \cdot 5$, $105 = 3 \cdot 5 \cdot 7$ and $150 = 2 \cdot 3 \cdot 5 \cdot 5$.

The common prime factors for all of them are 3 and 5. So, the GCF of these three numbers is $3 \cdot 5 = 15$.

6. Find the greatest common factor of the numbers.

a. 120 and 66	**b.** 36 and 136
c. 98 and 76	**d.** 132 and 72
e. 45 and 76	**f.** 64 and 120

7. Find the greatest common factor of the given numbers.

a. 75, 25 and 90	**b.** 54, 36 and 40
c. 18, 24 and 36	**d.** 72, 60 and 48

Find the greatest common factor of 187 and 264.

Puzzle Corner

Factoring Sums

1. You have seen these before! Write two different expressions for the total area, thinking of: (1) the area of the big rectangle as a whole and (2) the sum of the areas of the two small rectangles.

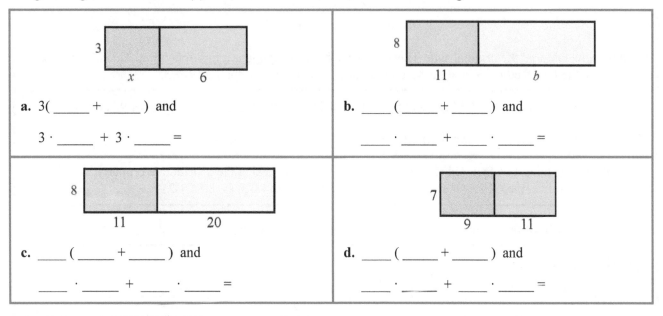

a. 3(_____ + _____) and

3 · _____ + 3 · _____ =

b. _____ (_____ + _____) and

_____ · _____ + _____ · _____ =

c. _____ (_____ + _____) and

_____ · _____ + _____ · _____ =

d. _____ (_____ + _____) and

_____ · _____ + _____ · _____ =

2. List the length and width of all the possible rectangles with an area of 30 cm^2 and where the sides are whole numbers (in centimeters). The first one would be 1 cm × 30 cm, and the second would be 2 cm × 15 cm.

3. List the length and width of all of the possible rectangles that have an area of 40 m^2 and where the sides are whole numbers (in meters). The first one would be 1 m × 40 m.

4. Build a big rectangle out of two smaller ones like in exercise #1. Choose one rectangle from exercise #2 and one from #3 that each have one side that is the same length. In the grid at the right, sketch the two rectangles that you chose side by side, touching so that they share the side that is the same length. Your sketch should look like the figures in exercise #1.

5. Can you find another answer to exercise 4?

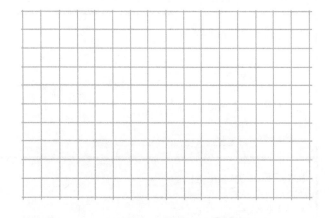

Another usage for the GCF is to factor expressions. **Factoring an expression** means writing it as a product (a multiplication).

Example 1. We can easily write the sum 54 + 27 as multiplication, once we notice that both 54 and 27 have the factor 9. So, 54 + 27 is $9 \cdot 6 + 9 \cdot 3$. Now, using the distributive property "backwards," we write

$$9 \cdot 6 + 9 \cdot 3 = 9(6 + 3)$$

We have now factored the original sum. This means we have written it as a multiplication. This time we have two factors: the first factor is 9, and the second factor is actually a *sum*: the sum 6 + 3.

Example 2. Write the sum 92 + 56 as a multiplication, using the greatest common factor of 92 and 56.

The GCF of 92 and 56 is 4. So, we write $92 + 56 = 4 \cdot 23 + 4 \cdot 14 = 4(23 + 14)$.

Example 3. You have also factored expressions such as $72x + 16$ using the distributive property "backwards." Notice, the GCF of 72 and 16 is 8. We can write $72x + 16$ as $8(9x + 2)$.

6. First find the GCF of the numbers. Then factor the expressions using the GCF.

a. GCF of 18 and 12 is 6

18 + 12 = 6 · 3 + 6 · 2 = 6 (___ + ___)

b. GCF of 6 and 10 is ___

6 + 10 = ___ · ___ + ___ · ___ = ___ (___ + ___)

c. GCF of 22 and 11 is ___

22 + 11 = ___ · ___ + ___ · ___ = ___ (___ + ___)

d. GCF of 15 and 21 is _____

15 + 21 = ___ · ___ + ___ · ___ = ___ (___ + ___)

e. GCF of 25 and 35 is _____

25 + 35 = ___ (___ + ___)

f. GCF of 72 and 86 is _____

72 + 86 = ___ (___ + ___)

g. GCF of 96 and 40 is _____

96 + 40 = ___ (___ + ___)

h. GCF of 39 and 81 is _____

39 + 81 = ___ (___ + ___)

7. **a.** Express the sum 32 + 40 as a product (multiplication).

   b. Draw two rectangles, side by side, to represent the product you wrote.

8. Draw two rectangles, side by side, to represent the sum 30 + 25.

9. Draw three rectangles, side by side, to represent the sum 42 + 24 + 30.

10. You need to build a rectangular animal pen that has an area of 45 m². If the lengths of the sides need to be in whole meters, what are the options?

11. **a.** List a pair of numbers whose greatest common factor is 1.

    b. List two more pairs of numbers whose greatest common factor is 1.

12. Write these sums as a product (multiplication) of their GCF and another sum.

a. The GCF of 15 and 5 is _____ $15x + 5 =$ ___ (___ + ___)	**b.** The GCF of 18 and 30 is _____ $18x + 30 =$ ___ (___ + ___)
c. The GCF of 72 and 54 is _____ $72a + 54b =$ ___ (___ + ___)	**d.** The GCF of 100 and 90 is _____ $100y + 90x =$ ___ (___ + ___)

> **Puzzle Corner**
>
> **a.** List two numbers whose greatest common factor is 13.
>
> **b.** List two numbers whose greatest common factor is 51.

The Least Common Multiple (LCM)

A **multiple** of a whole number n is any of the numbers n, $2n$, $3n$, $4n$, $5n$ and so on. In other words, a whole number times the number n is a multiple of n.

Example 1. The multiples of 9 are 9, 18, 27, 36, 45, 54, 63 and so on.

When we have two or more whole numbers, we can find their **least common multiple**.

As in the case of the greatest common factor, the term "least common multiple" itself tells us what it is! Read it again: least common multiple. All we need to do (in principle) is to find the multiples of the numbers, then find the common multiples, and lastly choose the one that is the least, or the smallest.

Example 2. Find the least common multiple of 5 and 8.
- The multiples of 5 are: 5, 10, 15, 20, 25, 30, 35, <u>40</u>, 45, 50, ... <u>80</u>,...
- The multiples of 8 are: 8, 16, 24, 32, <u>40</u>, 48, 56, 64, 72, <u>80</u>, ...

Among these multiples we find the common multiples 40 and 80. There are others as well, such as 120, 160 and so on, but 40 is the least (smallest) common multiple of 5 and 8.

Also, 40 is 5 · 8. Note: $a \cdot b$ is *always* a common multiple of both a and b, but it is not always the *least* common multiple.

Example 3. Find the least common multiple of 4 and 6.
- The multiples of 4 are: 4, 8, <u>12</u>, 16, 20, <u>24</u>, 28, 32, <u>36</u>, 40, ...
- The multiples of 6 are: 6, <u>12</u>, 18, <u>24</u>, 30, <u>36</u>, 42, 48, 54, 60, ...

Among these, we find the common multiples 12, 24 and 36. The *least* common multiple (LCM) is 12.

Note that the LCM of 4 and 6 is *not* 4 · 6.

1. Find the LCM of these numbers.

a. 2 and 6	**b.** 6 and 9
c. 14 and 8	**d.** 3 and 8
e. 7 and 10	**f.** 10 and 15

2. **a.** List four multiples of 6 that are less than 100.

   b. What is the biggest multiple of 4 that is less than 100?

   c. What is the smallest multiple of 250 that is more than 1,000?

3. A bus running Route A leaves the main bus terminal at every 15 minutes, and a bus running Route B at every 12 minutes. If there was a bus for both routes leaving at 3:30 PM, when is the next time that there is a bus for both routes leaving at the same time?

4. Boxes that are 20 cm tall are being stacked, as well as boxes that are 45 cm tall. What is the least height in centimeters at which both stacks are the same height?

5. Go back to exercise #1. Is the LCM of the two numbers also their product?

a. Is the LCM of 2 and 6 equal to 2 · 6?	**b.** Is the LCM of 6 and 9 equal to 6 · 9?
c. Is the LCM of 14 and 8 equal to 14 · 8?	**d.** Is the LCM of 3 and 8 equal to 3 · 8?
e. Is the LCM of 7 and 10 equal to 7 · 10?	**f.** Is the LCM of 10 and 15 equal to 10 · 15?

> You may be wondering why sometimes the LCM is the product of the two numbers, and other times it is not. The key is:
>
> If the numbers do not have *any common factors* (except 1), then their LCM is their product.
>
> **Example 4.** The numbers 8 and 10 have a common factor 2. Therefore, their LCM will not be 80. (Can you find what it is?)

6. Find the LCM of these numbers.

a. 3 and 4	**b.** 9 and 7
c. 10 and 5	**d.** 4 and 7
e. 2 and 10	**f.** 4 and 10

7. **a.** Draw a line from each number to the correct box.

    b. Which number is a "black sheep" (neither a factor nor a multiple of 24)?

    c. Which number is BOTH a factor and a multiple of 24?

240 8 48 4 96 24 1 2

| a factor of 24 | a multiple of 24 |

120 3 30 72 144 6 12

Example 5. Find the least common multiple of 4, 6, and 5.

To find the LCM of three numbers, you could make three lists of their multiples, but a quicker way is to *first* find the LCM of two of the numbers, and then use that.

The LCM of 4 and 6 is 12. Now we will find the LCM of 12 and 5, and that will be the LCM of 4, 6, and 5.

Since 12 and 5 don't have common factors, their LCM is $12 \cdot 5 = 60$, and that is also the LCM of 4, 6, and 5.

8. Find the LCM of three numbers.

a. 3 and 8 and 6	**b.** 2 and 6 and 10
c. 3 and 5 and 2	**d.** 4 and 7 and 8

9. Anne is a hair stylist. Among her customers, Mrs. Goodwill comes for a haircut every 8 weeks, Ms. Sidney every 6 weeks, and Ms. Locksmith every 4 weeks. If all of them came for a haircut on a certain Monday, in how many weeks will they again have a haircut on the same day?

Remember? Before adding or subtracting *unlike* fractions, we need to convert them into equivalent fractions that have a **common denominator**.

Example 6 (on the right). The denominators 8 and 10 need to "go into" the common denominator. In other words, the common denominator must be **a multiple of both** 8 and 10. Naturally, the least common multiple is what we are looking for! The LCM of 8 and 10 is 40.

You *could* use any common multiple as the common denominator, (such as 80), but the LCM is the least (smallest) common denominator.

$$\frac{5}{8} + \frac{1}{10}$$
$$\downarrow \qquad \downarrow$$
$$\frac{25}{40} + \frac{4}{40} = \frac{29}{40}$$

10. Add or subtract the fractions. Give your answer in lowest terms.

a. $\dfrac{1}{9} + \dfrac{1}{8}$	**b.** $\dfrac{1}{12} + \dfrac{7}{8}$
c. $\dfrac{3}{7} - \dfrac{3}{10}$	**d.** $\dfrac{8}{9} - \dfrac{1}{6}$

(*This section is optional.*) There is a **quicker and more efficient way for finding the least common multiple of numbers**, where we do not have to make lists of multiples. It is based on prime factorization.

1. Write the prime factorization of the numbers.
2. The LCM is formed by taking ALL the factors from the numbers, without repeating any common factor.

Example 7. Find the LCM of 45 and 25.

1. The prime factorizations are $45 = 3 \cdot 3 \cdot 5$ and $25 = 5 \cdot 5$.

2. Form a number that is "all inclusive" or that includes all the factors from both numbers. It is $3 \cdot 3 \cdot 5 \cdot 5$, which equals 225.

Notice that $3 \cdot 3 \cdot 5 \cdot 5$ includes *both* $3 \cdot 3 \cdot 5$ and $5 \cdot 5$ but has no other factors beyond those.

Example 8. Find the LCM of 24 and 40.

1. The prime factorizations are $24 = 2 \cdot 2 \cdot 2 \cdot 3$ and $40 = 2 \cdot 2 \cdot 2 \cdot 5$.

2. The number that includes all the factors from both numbers is $2 \cdot 2 \cdot 2 \cdot 3 \cdot 5 = 120$.

11. Find the least common multiple of the numbers using any method.

a. 40 and 15	**b.** 20 and 24
c. 20 and 16	**d.** 50 and 120

12. Convert the fractions so they have the same denominator, and then compare them. Your work above can help!

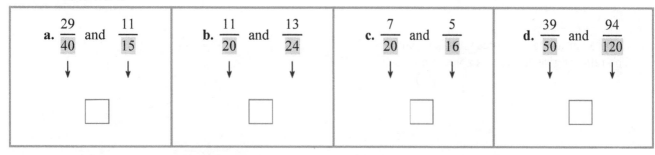

a. $\dfrac{29}{40}$ and $\dfrac{11}{15}$ b. $\dfrac{11}{20}$ and $\dfrac{13}{24}$ c. $\dfrac{7}{20}$ and $\dfrac{5}{16}$ d. $\dfrac{39}{50}$ and $\dfrac{94}{120}$

Puzzle Corner

If the first day of the year is Tuesday, what day of the week is day number 236?

Chapter 6 Mixed Review

1. Solve. (Powers and Exponents/Ch.1)

| a. $10^4 \cdot 3$ | b. 7^3 | c. $10 \cdot 5^3$ |

2. Write in expanded form using exponents. (Powers and Exponents/Ch.1)

   a. 109,200

   b. 7,002,050

3. Andrew cut a 9-foot board into two pieces that are in a ratio of 3:5.
   Find the length of each of the two pieces.
   (Ratio Problems and Bar Models 1/Ch.4)

4. Convert each division problem into another, equivalent division problem that you can solve in your head.
   (Review: Divide Decimals by Decimals/Ch.3)

| a. $\dfrac{16}{0.4} = \underline{} =$ | b. $\dfrac{7}{0.007} = \underline{} =$ | c. $\dfrac{99}{0.11} = \underline{} =$ |

5. Multiply. (Review: Multiply and Divide Decimals Mentally/Ch.3)

| a. $100 \cdot 0.2 =$ _____ | b. $3 \cdot 1.02 =$ _____ | c. $0.9 \cdot 0.2 \cdot 0.5 =$ _____ |
| $120 \cdot 0.02 =$ _____ | $5 \cdot 3.02 =$ _____ | $30 \cdot 0.005 \cdot 0.2 =$ _____ |

6. a. Draw a bar model to represent this situation: (Ratio Problems and Bar Models 1/Ch.4)
   The ratio of girls to boys in a vocational school is 7:4.

   b. What is the ratio of boys to all of the students?

   c. If there are 748 students in all, how many are girls?
      How many are boys?

7. Liz is 150 cm tall, and her dad is 1.8 m tall. What
   percentage is Liz's height of her dad's height?
   (Percent/Ch.5)

8. The Madison family spent $540 for groceries in one month. That was 24% of their total budget.
   How much was their total budget for the month?
   (Finding the Total When the Percentage Is Known/Ch.5)

9. Which is cheaper, a $180 camera discounted by 20%,
   or a $155 camera discounted by 10%?
   (Discounts/Ch.5)

10. Multiply using the distributive property. (The Distributive Property/Ch.2)

a. $2(7m + 4) =$	b. $10(x + 6 + 2y) =$

11. Write an expression. (Words and Expressions/Ch.2)

   a. the quantity $5s$ plus 8, divided by 7

   b. the quantity n plus 11, cubed

   c. y more than 8

   d. x divided by y squared

12. Solve the inequality $x - 3 < 0$ in the set $\{-2, -1, 0, 1, 2, 3\}$.
    (Inequalities/Ch.2)

13. Divide, giving your answer as a decimal. If necessary, round the answers to three decimal digits.
    (Review: Divide Decimals by Decimals and Fractions and Decimals/Ch.2)

a. $0.928 \div 0.3$	b. $\dfrac{7}{34}$

Chapter 6 Review

1. Factor the following composite numbers into their prime factors.

a. 81 /\	b. 26 /\	c. 65 /\
d. 96 /\	e. 124 /\	f. 450 /\

2. Simplify.

a. $\dfrac{28}{84} = \dfrac{4 \cdot 7}{21 \cdot 4} =$	b. $\dfrac{75}{160} =$
c. $\dfrac{222}{36} =$	d. $\dfrac{48}{120} =$

3. Find the least common multiple of these pairs of numbers.

a. 3 and 7	b. 10 and 8
c. 11 and 6	d. 6 and 8

4. Find the greatest common factor of the given number pairs.

a. 24 and 64	b. 100 and 75
c. 80 and 96	d. 78 and 96

5. Fill in with the words "multiple(s)" or "factor(s)."

   a.
   - 25, 50, 75, 100, 125 and 150 are _____ of 25.
   - 1, 2, 5, 10, 25 and 50 are _____ of 50.
   - Each number has an infinite number of _____.
   - Each number has a greatest _____.
   - If a number x divides into another number y, we say x is a _____ of y.

   b. List five different multiples of 15 that are less than 200 but more than 60.

   c. Find five numbers that are multiples of both 4 and 7.

   What is the LCM of 4 and 7?

6. First, find the GCF of the numbers. Then factor the expressions using the GCF.

a. GCF of 12 and 21 is _____   12 + 21 = ____ · ____ + ____ · ____ = ____ (____ + ____)
b. GCF of 45 and 70 is _____   45 + 70 = ____ (____ + ____)

7. Draw two rectangles, side by side, to represent the sum 42 + 30.

Chapter 7: Fractions
Introduction

This chapter begins with a review of fraction arithmetic from fifth grade—specifically, addition, subtraction, simplification, and multiplication of fractions. Then it focuses on division of fractions.

The introductory lesson on the division of fractions presents the concept of reciprocal numbers and ties the reciprocity relationship to the idea that division is the appropriate operation to solve questions of the form, "How many times does this number fit into that number?" For example, we can write a division from the question, "How many times does 1/3 fit into 1?" The answer is, obviously, 3 times. So we can write the division $1 \div (1/3) = 3$ and the multiplication $3 \cdot (1/3) = 1$. These two numbers, 3/1 and 1/3, are reciprocal numbers because their product is 1.

Students learn to solve questions like that through using visual models and writing division sentences that match them. Thinking of fitting the divisor into the dividend (measurement division) also gives us a tool to check whether the answer to a division problem is reasonable.

Naturally, the lessons also present the shortcut for fraction division—that each division can be changed into a multiplication by taking the reciprocal of the divisor, which is often called the "invert (flip)-and-multiply" rule. However, that "rule" is just a shortcut. It is necessary to memorize it, but memorizing a shortcut doesn't help students make sense conceptually out of the division of fractions—they also need to study the concept of division and use visual models to better understand the process involved.

In two lessons that follow, students apply what they have learned to solve problems involving fractions or fractional parts. A lot of the problems in these lessons are review in the sense that they involve previously learned concepts and are similar to problems students have solved earlier, but many involve the division of fractions, thus incorporating the new concept presented in this chapter.

Consider mixing the lessons from this chapter (or from some other chapter) with the lessons from the geometry chapter (which is a fairly long chapter). For example, the student could study these topics and geometry on alternate days, or study a little from both each day. Such, somewhat spiral, usage of the curriculum can help prevent boredom, and also to help students retain the concepts better.

Also, don't forget to use the resources for challenging problems:
https://l.mathmammoth.com/challengingproblems
I recommend that you at least use the first resource listed, Math Stars Newsletters.

The Lessons in Chapter 7

	page	span
Review: Add and Subtract Fractions and Mixed Numbers	37	*4 pages*
Add and Subtract Fractions: More Practice	41	*3 pages*
Review: Multiplying Fractions 1	44	*3 pages*
Review: Multiplying Fractions 2	47	*3 pages*
Dividing Fractions: Reciprocal Numbers	50	*5 pages*
Divide Fractions	55	*4 pages*
Problem Solving with Fractions 1	59	*3 pages*
Problem Solving with Fractions 2	62	*3 pages*
Chapter 7 Mixed Review	65	*2 pages*
Fractions Review	67	*3 pages*

Helpful Resources on the Internet

We have compiled a list of Internet resources that match the topics in this chapter. This list of links includes web pages that offer:

- **online practice** for concepts;
- online **games**, or occasionally, printable games;
- **animations** and interactive **illustrations** of math concepts;
- **articles** that teach a math concept.

We heartily recommend you take a look at the list. Many of our customers love using these resources to supplement the bookwork. You can use the resources as you see fit for extra practice, to illustrate a concept better, and even just for some fun. Enjoy!

https://l.mathmammoth.com/gr6ch7

Review: Add and Subtract Fractions and Mixed Numbers

Example 1. Add $\dfrac{5}{6} + 2\dfrac{5}{8}$.

We need to convert unlike fractions into equivalent fractions that have a common denominator before we can add them. The common denominator must be a **multiple of both 6 and 8** (a *common* multiple).

Naturally, $6 \cdot 8 = 48$ is one common multiple of 6 and 8. We could use 48. However, it is better to use 24, which is the *least* common multiple (LCM) of 6 and 8, because it leads to easier calculations.

The common denominator is 24:

$$\dfrac{5}{6} + 2\dfrac{5}{8}$$
$$\downarrow \qquad \downarrow$$
$$\dfrac{20}{24} + 2\dfrac{15}{24} = 2\dfrac{35}{24} = 3\dfrac{11}{24}$$

1. Write the addition sentences.

a. ___ + ___ = ___ + ___ = ___

b. $\dfrac{3}{4} + \dfrac{1}{9}$ → ___ + ___ = ___

c. $\dfrac{7}{10} + \dfrac{1}{4}$ → ___ + ___ = ___

2. Add and subtract. Use the common denominator you found in the previous exercise. Remember, the *best* possible choice for the common denominator (but not the only one) is the LCM of the denominators.

a. $\dfrac{5}{16} + \dfrac{1}{6}$	b. $3\dfrac{1}{12} + 1\dfrac{4}{9}$	c. $\dfrac{5}{6} - \dfrac{3}{8}$
d. $2\dfrac{5}{12} + \dfrac{4}{5}$	e. $5\dfrac{11}{15} - 2\dfrac{3}{20}$	f. $\dfrac{45}{100} + \dfrac{9}{20}$

| **Regroup in subtraction**, if necessary. Here we regroup **one** as 13/13. This leaves 9 wholes. There is already 1/13 in the column of the fractional parts, so in total we get 14/13. | $9\frac{14}{13}$ $\cancel{10}\frac{\cancel{1}}{13}$ $-5\frac{5}{13}$ $\rule{2cm}{0.4pt}$ $4\frac{9}{13}$ | **We can use the same idea (regrouping) when the fractions are written horizontally.** Take one of the 7 wholes, think of it as 9/9, and regroup that with the fractional parts (with 2/9). Instead of 7 wholes, we are left with 6, and instead of 2/9, we get 11/9. | $7\frac{2}{9} - 3\frac{8}{9}$ $\downarrow \downarrow$ $6\frac{11}{9} - 3\frac{8}{9} = 3\frac{3}{9}$ |

3. Subtract.

a. $7\frac{3}{9}$ $-2\frac{7}{9}$

b. $18\frac{1}{10}$ $-5\frac{9}{10}$

c. $10\frac{1}{15}$ $-3\frac{8}{15}$

d. $16\frac{3}{9} - 9\frac{8}{9}$

e. $7\frac{3}{14} - 2\frac{10}{14}$

4. Subtract. First write equivalent fractions with the same denominator.

a. $3\frac{3}{4} \rightarrow 3\frac{}{}$
   $-1\frac{1}{6} \rightarrow -1\frac{}{}$

b. $3\frac{3}{8} \rightarrow$
   $-1\frac{5}{12} \rightarrow -$

c. $8\frac{9}{11} \rightarrow$
   $-5\frac{1}{2} \rightarrow -$

5. Figure out and explain how these subtractions were done!

Emma's way: $9\frac{2}{17} - 3\frac{8}{17}$

$= (9 - 3) + (\frac{2}{17} - \frac{8}{17}) = 6 - \frac{6}{17} = 5\frac{11}{17}$

Joe's method: $5\frac{3}{14} - 2\frac{9}{14}$
$\downarrow$
$5\frac{3}{14} - 2\frac{3}{14} - \frac{6}{14}$

$= 3 - \frac{6}{14} = 2\frac{8}{14}$

When adding or subtracting three or more fractions, find a common denominator for all of them. You can always use the product of the denominators as your common denominator. However, it *may be* more efficient to use the LCM of the denominators if it is smaller.

Example 2. Here, we *could* use 6 · 7 · 2 = 84 as a common denominator.

However, in this case, the LCM of 6, 7 and 2 is 42, so it is better (leads to easier calculations) than using 84.

Another option would be to add the first two fractions (5/6 and 5/7) to get 65/42, and then to subtract the third fraction, 1/2, from that result.

$$\frac{5}{6} + \frac{5}{7} - \frac{1}{2}$$
$$\downarrow \quad \downarrow \quad \downarrow$$
$$\frac{35}{42} + \frac{30}{42} - \frac{21}{42} = \frac{44}{42} = 1\frac{1}{21}$$

6. Add or subtract the fractions.

a. $\dfrac{5}{12} + \dfrac{1}{6} + \dfrac{1}{3}$	**b.** $\dfrac{2}{7} + \dfrac{1}{2} - \dfrac{1}{4}$
c. $\dfrac{1}{10} + \dfrac{2}{5} + \dfrac{1}{3}$	**d.** $\dfrac{19}{20} - \dfrac{1}{3} - \dfrac{1}{4}$
e. $\dfrac{7}{8} - \dfrac{1}{5} + \dfrac{2}{3}$	**f.** $\dfrac{7}{6} - \dfrac{3}{5} + \dfrac{3}{4}$

7. Joe started working at an automobile company 23 ½ years ago. However, during that time, he has taken ¼ of a year off for paternity leave, and spent another 1 ⅓ years laid off due to a recession. So, how long has he actually been working for the company?

While you can often compare two fractions using mental math strategies, sometimes the fractions are so close to each other that you need to rewrite both using a common denominator, then compare.

Example 3. Which is more, $\frac{7}{8}$ or $\frac{11}{13}$?

Let's write both using the common denominator 104 (on the right).

We see that 7/8 is more.

$$\frac{7}{8} \quad \frac{11}{13}$$
$$\downarrow \quad \downarrow$$
$$\frac{91}{104} > \frac{88}{104}$$

8. Compare the fractions, writing < or > between them. Use a common denominator only if you need to.

a. $\frac{1}{2} \square \frac{5}{9}$ b. $\frac{15}{65} \square \frac{15}{34}$ c. $\frac{6}{15} \square \frac{1}{2}$ d. $\frac{1}{120} \square \frac{1}{75}$

e. $\frac{3}{5} \square \frac{8}{13}$ f. $\frac{2}{3} \square \frac{8}{11}$ g. $\frac{11}{15} \square \frac{3}{4}$ h. $\frac{10}{2{,}000} \square \frac{2}{1{,}000}$

9. Julie is convinced that 5/6 is more than 7/8 — she even sketched a picture where it looks like it is so.

   How would you convince (prove to) her otherwise?

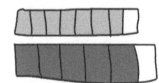

10. Order the fractions from the smallest to the biggest.

a. $\frac{1}{4}, \frac{1}{2}, \frac{3}{8}, \frac{3}{7}$

___ < ___ < ___ < ___

b. $\frac{2}{3}, \frac{7}{5}, \frac{5}{4}, \frac{3}{2}$

___ < ___ < ___ < ___

c. $\frac{2}{3}, \frac{8}{5}, \frac{3}{5}, \frac{3}{4}, \frac{5}{4}$

___ < ___ < ___ < ___

d. $\frac{5}{6}, \frac{7}{12}, \frac{5}{8}, \frac{2}{9}, \frac{6}{5}$

___ < ___ < ___ < ___

Puzzle Corner

Solve the equations. *Hint:* if the fractions confuse you, *first* think how the equation would be solved if it had whole numbers. Then solve the original equation the same way.

a. $8\frac{4}{7} + x = 10\frac{2}{5}$

b. $5\frac{1}{9} - x = 2\frac{1}{3}$

Add and Subtract Fractions: More Practice

These exercises simply give you more practice on adding and subtracting fractions and mixed numbers. Use them as directed by your teacher.

1. Add or subtract. Give your answer in lowest terms, and as a mixed number, if applicable.

a. $\dfrac{17}{18} + \dfrac{2}{9}$	b. $\dfrac{11}{30} + \dfrac{7}{12}$	c. $\dfrac{13}{22} + \dfrac{3}{4}$
d. $6\dfrac{7}{10} - 1\dfrac{3}{20}$	e. $4\dfrac{7}{8} - 1\dfrac{1}{3}$	f. $15\dfrac{9}{10} - 3\dfrac{31}{100}$

2. Subtract. First write equivalent fractions with the same denominator.

a. $5\dfrac{1}{2} \to 5\dfrac{}{}$ $-\ 1\dfrac{7}{12} \to -\ 1\dfrac{}{}$	b. $12\dfrac{1}{9}$ $-\ 5\dfrac{2}{3}$	c. $33\dfrac{1}{3}$ $-\ 17\dfrac{6}{7}$
d. $8\dfrac{1}{9}$ $-\ 2\dfrac{7}{12}$	e. $86\dfrac{6}{7}$ $-\ 45\dfrac{1}{8}$	f. $53\dfrac{1}{6}$ $-\ 40\dfrac{6}{7}$

3. Add or subtract these fractions to solve the riddle! Give your answer as a mixed number, if possible, and put the fractional part into lowest terms.

I. $2\frac{7}{15} + 1\frac{1}{3}$

P. $\frac{4}{7} - \frac{1}{5}$

V. $6\frac{1}{15} + 5\frac{1}{2}$

I. $9\frac{7}{20} - 3\frac{2}{10}$

U. $2\frac{1}{8} - \frac{1}{6}$

T. $9\frac{1}{12} - 2\frac{9}{30}$

D. $3\frac{7}{12} - 2\frac{1}{3}$

G. $9\frac{1}{10} - 4\frac{3}{4}$

E. $5\frac{5}{14} - 3\frac{1}{4}$

D. $\frac{5}{22} + 2\frac{1}{11}$

L. $4\frac{5}{9} - 1\frac{1}{12}$

M. $2\frac{1}{4} - \frac{1}{3}$

I. $2\frac{4}{5} + \frac{11}{24}$

I. $3\frac{2}{7} - 2\frac{1}{5}$

I. $5\frac{2}{9} - \frac{11}{27}$

D. $6\frac{3}{20} - 3\frac{1}{12}$

N. $\frac{3}{8} - \frac{1}{12}$

L. $4\frac{7}{8} - 1\frac{1}{12}$

Why did the amoeba flunk the math test? Because it...

$1\frac{11}{12}$	$1\frac{23}{24}$	$3\frac{17}{36}$	$6\frac{47}{60}$	$3\frac{4}{5}$	$\frac{13}{35}$	$3\frac{19}{24}$	$3\frac{31}{120}$	$2\frac{3}{28}$	$1\frac{1}{4}$

by

$2\frac{7}{22}$	$6\frac{3}{20}$	$11\frac{17}{30}$	$1\frac{3}{35}$	$3\frac{1}{15}$	$4\frac{22}{27}$	$\frac{7}{24}$	$4\frac{7}{20}$

4. Solve the expressions. Put each answer into lowest terms. Then find it in the grid. Color the squares of the grid that contain the answers in a bright color. Color the other squares in a dark color. Note the pattern.

a. $3\dfrac{1}{3} - 1\dfrac{2}{15} + 1\dfrac{2}{5}$

$5\dfrac{1}{48}$	$2\dfrac{7}{24}$	$2\dfrac{5}{48}$	$5\dfrac{3}{50}$	$1\dfrac{1}{60}$
$3\dfrac{11}{48}$	$6\dfrac{1}{8}$	$4\dfrac{5}{24}$	$3\dfrac{17}{24}$	$2\dfrac{1}{50}$
$3\dfrac{53}{66}$	$2\dfrac{19}{48}$	$2\dfrac{9}{50}$	$4\dfrac{1}{8}$	$3\dfrac{7}{8}$
$3\dfrac{1}{5}$	$3\dfrac{13}{15}$	$3\dfrac{3}{5}$	$5\dfrac{13}{66}$	$5\dfrac{4}{5}$
$4\dfrac{53}{66}$	$1\dfrac{29}{60}$	$4\dfrac{11}{50}$	$7\dfrac{39}{50}$	$6\dfrac{41}{50}$

b. $\dfrac{7}{10} + \dfrac{2}{25} + 1\dfrac{2}{5}$

c. $6\dfrac{67}{100} - 1\dfrac{2}{5} + 1\dfrac{11}{20}$

d. $3\dfrac{1}{2} - \dfrac{2}{3} - \dfrac{7}{16}$

g. $3\dfrac{7}{20} - 1\dfrac{1}{12} - 1\dfrac{1}{4}$

e. $\dfrac{13}{16} + 2\dfrac{1}{12} + 2\dfrac{3}{24}$

h. $5\dfrac{1}{6} + 1\dfrac{3}{8} - 2\dfrac{1}{3}$

f. $7\dfrac{7}{8} - 1\dfrac{1}{2} - 2\dfrac{1}{4}$

i. $19\dfrac{7}{11} - 10\dfrac{1}{3} - 4\dfrac{1}{2}$

Review: Multiplying Fractions 1

The **shortcut for multiplying fractions** is: • Multiply the numerators. • Multiply the denominators.	$\dfrac{6}{7} \cdot \dfrac{5}{2} \cdot \dfrac{1}{3} = \dfrac{6 \cdot 5 \cdot 1}{7 \cdot 2 \cdot 3} = \dfrac{30}{42} = \dfrac{5}{7}$
To **multiply mixed numbers**, *first* write them as fractions, then multiply. $2\dfrac{1}{3} \cdot 1\dfrac{1}{10} = \dfrac{7}{3} \cdot \dfrac{11}{10} = \dfrac{7 \cdot 11}{3 \cdot 10} = \dfrac{77}{30} = 2\dfrac{17}{30}$	If one of the factors is a whole number, write it as a fraction with a denominator of 1. $6 \cdot \dfrac{11}{12} = \dfrac{6}{1} \cdot \dfrac{11}{12} = \dfrac{66}{12} = \dfrac{11}{2} = 5\dfrac{1}{2}$

1. Multiply. Give your answer in lowest terms, and as a mixed number, if applicable.

a. $5 \cdot \dfrac{7}{8}$	**b.** $\dfrac{2}{7} \cdot \dfrac{5}{6}$
c. $\dfrac{9}{10} \cdot \dfrac{6}{7} \cdot \dfrac{1}{2}$	**d.** $1\dfrac{1}{3} \cdot 2\dfrac{2}{3}$
e. $\dfrac{1}{10} \cdot 3\dfrac{1}{5}$	**f.** $2\dfrac{5}{6} \cdot 10 \cdot \dfrac{1}{2}$

2. Find the area of a square with sides 1 ¼ units. Use fractions.

3. A biscuit recipe calls for 1½ cups of buttermilk, and Mary plans to make the recipe one and a half times, three times a week, in order to sell biscuits. How much buttermilk does she need in a week?

You have already learned to use **factoring** when simplifying.

The example on the right shows simplifying 96/144.

$$\frac{96}{144} = \frac{\overset{2}{\cancel{8}} \cdot \overset{1}{\cancel{12}}}{\underset{3}{\cancel{12}} \cdot \underset{1}{\cancel{12}}} = \frac{2}{3}$$

You have also learned how to simplify **"criss-cross."** To simplify 45/150, we cancel the 5s from the numerator and the denominator. Then we simplify 9 and 30 into 3 and 10.

$$\frac{45}{150} = \frac{\overset{1}{\cancel{5}} \cdot \overset{3}{\cancel{9}}}{\underset{10}{\cancel{30}} \cdot \underset{1}{\cancel{5}}} = \frac{3}{10}$$

In a similar manner, you can simplify fractions *before* multiplying.

Compare the two examples on the right. They show the same problem.

$$\frac{7}{6} \cdot \frac{3}{9} = \frac{7 \cdot \overset{1}{\cancel{3}}}{\underset{2}{\cancel{6}} \cdot 9} = \frac{7}{18}$$

The first one (above right) is written out with an extra step, whereas the one below is written without the extra step. In both cases, the simplifying is done *before* multiplying.

$$\frac{7}{\underset{2}{\cancel{6}}} \cdot \frac{\overset{1}{\cancel{3}}}{9} = \frac{7}{18}$$

4. Simplify before multiplying, and solve the riddle.

E. $\dfrac{3}{10} \cdot \dfrac{1}{3} =$

A. $\dfrac{5}{6} \cdot \dfrac{2}{4} =$

O. $\dfrac{2}{6} \cdot \dfrac{5}{7} =$

L. $\dfrac{2}{9} \cdot \dfrac{9}{11} =$

M. $\dfrac{4}{10} \cdot \dfrac{1}{3} =$

E. $\dfrac{3}{10} \cdot \dfrac{3}{9} =$

I. $7 \cdot \dfrac{5}{21} =$

N. $\dfrac{16}{24} \cdot 8 =$

W. $\dfrac{4}{5} \cdot \dfrac{3}{6} =$

P. $\dfrac{4}{8} \cdot \dfrac{1}{3} =$

S. $\dfrac{7}{40} \cdot 15 =$

R. $\dfrac{2}{6} \cdot \dfrac{3}{9} =$

| $\frac{5}{12}$ | $\frac{1}{9}$ | $\frac{1}{10}$ | $\frac{21}{8}$ | $\frac{5}{3}$ | $\frac{2}{15}$ | $\frac{1}{6}$ | $\frac{2}{11}$ | $\frac{1}{10}$ | $\frac{16}{3}$ | $\frac{5}{21}$ | $\frac{2}{5}$ |

These problems ☐ ☐ ☐ ☐ ☐ ☐ ☐ ☐ ☐ ☐ ☐ !

Use multiplication to find a fractional part of a fraction. The word "of" translates into multiplication.

How much is $\frac{3}{4}$ of ◯ ?

Since "of" becomes ·, we get the multiplication

$$\frac{3}{4} \cdot \frac{8}{12} = \frac{\cancel{3}^1}{\cancel{4}_1} \cdot \frac{\cancel{8}^2}{\cancel{12}_4} = \frac{2}{4} = \frac{1}{2}.$$

But, how can we make sense of that answer 1/2?

If you have 8 slices of a pie that was originally cut into twelfths, and you take 3/4 *of* those 8 slices, you will end up with 6 slices (of the original 12). And 6/12 is 1/2.

$\frac{3}{4}$ of ◯ is ◯.

5. The pictures show how much pizza is left. Find the given part of it. Write a multiplication sentence.

a. Find $\frac{1}{2}$ of ◯

$\frac{1}{2} \cdot \underline{} = $

b. Find $\frac{2}{3}$ of ◯

$\underline{} \cdot \underline{} = $

c. Find $\frac{1}{4}$ of ◯

$\underline{} \cdot \underline{} = $

d. Find $\frac{9}{10}$ of ◯

$\underline{} \cdot \underline{} = $

e. Find $\frac{1}{6}$ of ◯

$\underline{} \cdot \underline{} = $

f. Find $\frac{3}{8}$ of ◯

$\underline{} \cdot \underline{} = $

6. Rewrite the ingredients for the pancake recipe as 3/4 of the original amounts.

Pancakes

1 3/4 c milk
2 eggs
2 c flour
2 1/2 tsp baking powder
1/2 tsp salt
1 tsp cinnamon

Pancakes

_____ c milk
_____ eggs
_____ c flour
_____ tsp baking powder
_____ tsp salt
_____ tsp cinnamon

7. Isabella was riding her bicycle from her house to her friend's, which was 3/4 mile away. Then, 2/3 of the way there, she realized that she had forgotten something, so she had to return home. What distance did Isabella ride her bicycle from her home to the point where she turned back and then home again? Calculate the distance in two ways:

a. Using fractions.

b. Using decimals.

Review: Multiplying Fractions 2

Fraction multiplication and area

Study the picture. The *colored* rectangle is a fraction of the one square meter. Its top measures 1/3 m, and its side measures 3/4 m. To find its area, we multiply those fractions: $\frac{1}{3}$ m $\cdot$ $\frac{3}{4}$ m = $\frac{3}{12}$ m² = $\frac{1}{4}$ m².

The whole square is 1 m². So the colored rectangle is 3/12 = 1/4 of that area, or 1/4 m².

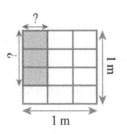

1. Write the multiplication for the area of the colored rectangle. In (c) and (d), do not include any units in the multiplication (simply write the fractions without any units).

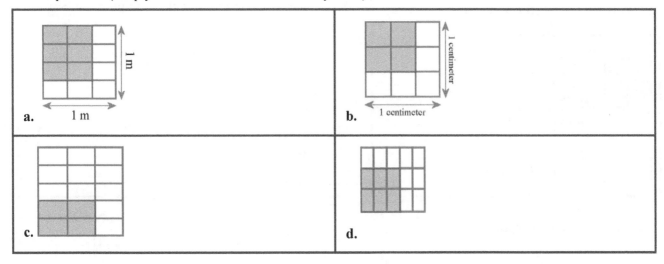

2. Multiply. Shade a rectangle in the grid to illustrate the multiplications.

a. $\frac{3}{4} \cdot \frac{3}{4} =$

b. $\frac{2}{3} \cdot \frac{3}{5} =$

c. $\frac{8}{9} \cdot \frac{1}{3} =$

3. Make equivalent fractions by multiplying the given fraction by different forms of the number 1.

a. Multiply the fraction by $\frac{5}{5}$.	b. Multiply the fraction by $\frac{3}{3}$.	c. Multiply the fraction by $\frac{9}{9}$.
___ $\cdot \frac{2}{3} =$	___ $\cdot \frac{7}{10} =$	___ $\cdot \frac{8}{15} =$

4. Is the result of multiplication more, less, or equal to the original number? Write <, >, or =. You do not have to calculate anything.

a. $\frac{11}{12} \cdot 21$ ☐ 21

b. $2\frac{1}{3} \cdot 19$ ☐ 19

c. $\frac{16}{16} \cdot 105$ ☐ 105

47

Multiplying mixed numbers - an area illustration

Study the picture carefully. The *colored* rectangle illustrates 1 2/3 · 1 2/3.

In this illustration, the sides of each *little* square are 1/3 units, with an area of 1/9 square unit, and each 3 · 3 square illustrates <u>one whole</u>.

The colored rectangle consists of 5 · 5 = 15 little squares.

This therefore equals 15 · 1/9 = 15/9 = 1 6/9 square units.

We get the same by multiplying the side lengths:

$1\frac{2}{3} \cdot 1\frac{2}{3} = \frac{5}{3} \cdot \frac{5}{3} = \frac{15}{9} = 1\frac{6}{9} = 1\frac{2}{3}$ square units.

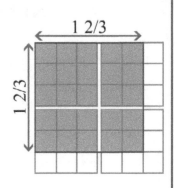

5. These situations use mixed numbers.

a. Write a multiplication and solve.

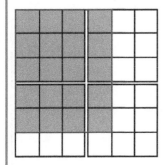

b. Shade a rectangle to illustrate the multiplication. Solve.

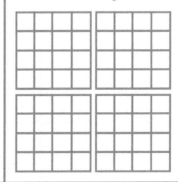

$1\frac{1}{4} \cdot 1\frac{3}{4} =$

6. Shade a rectangle to illustrate $3\frac{1}{3} \cdot 1\frac{2}{3}$.

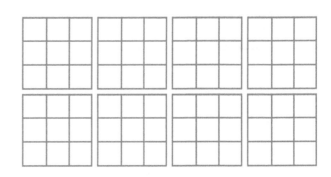

7. Leo, a 6th grade student, says: "Well, $3\frac{1}{3} \cdot 1\frac{2}{3}$ is simply 3 · 1 = 3, plus $\frac{1}{3} \cdot \frac{2}{3} = \frac{2}{9}$, a total of 3 2/9."

Use your drawing from (6) to explain why Leo's work is wrong.

8. A town covers over a 4 1/4 km by 3 3/8 km rectangle.

   a. Find its area using fractions.

   b. Now find its area using decimals, rounded to the nearest hundredth of a square kilometer.

You can simplify several times before multiplying.			
$\dfrac{\cancel{3}^{\,1}}{15} \cdot \dfrac{5}{\cancel{6}_{\,2}}$	$\dfrac{\cancel{3}^{\,1}}{\cancel{15}_{\,3}} \cdot \dfrac{\cancel{5}^{\,1}}{\cancel{6}_{\,2}} = \dfrac{1}{6}$	$\dfrac{\cancel{3}^{\,1}}{\cancel{15}_{\,5}} \cdot \dfrac{7}{14}$	$\dfrac{\cancel{3}^{\,1}}{\cancel{15}_{\,5}} \cdot \dfrac{\cancel{7}^{\,1}}{\cancel{14}_{\,2}} = \dfrac{1}{10}$
First simplify 3 and 6 into 1 and 2.	Then simplify 5 and 15 into 1 and 3.	First simplify 3 and 15 into 1 and 5.	Then simplify 7 and 14 into 1 and 2.

9. Simplify before you multiply.

a. $\dfrac{8}{12} \cdot \dfrac{6}{12}$	b. $\dfrac{3}{10} \cdot \dfrac{2}{18}$	c. $\dfrac{2}{30} \cdot \dfrac{10}{11}$
d. $\dfrac{7}{21} \cdot \dfrac{3}{4}$	e. $\dfrac{2}{16} \cdot \dfrac{8}{9}$	f. $\dfrac{18}{24} \cdot \dfrac{8}{9}$

10. Find the price for 2 3/4 pounds of nuts if one pound costs $8.

11. Find the price for 5/8 of a pound of nuts if one pound costs $10.

12. Try your simplifying skills with multiplying three fractions.

a. $\dfrac{5}{4} \cdot \dfrac{12}{9} \cdot \dfrac{3}{15}$	b. $\dfrac{8}{10} \cdot \dfrac{15}{27} \cdot \dfrac{9}{16}$
c. $\dfrac{1}{18} \cdot \dfrac{24}{33} \cdot \dfrac{9}{20}$	d. $\dfrac{3}{5} \cdot \dfrac{15}{18} \cdot \dfrac{16}{50}$

Puzzle Corner Simplify: $\dfrac{60}{48} \cdot \dfrac{36}{90} =$

Dividing Fractions: Reciprocal Numbers

One interpretation of division is **measurement division**, where we think: *How many times does one number go into another?* For example, to solve how many times 11 fits into 189, we divide 187 ÷ 11 = 17.

(The other interpretation is equal sharing; we will come to that later.)

Let's apply that to fractions. How many times does go into ?

We can solve this just by looking at the pictures: three times. We can write the division: $2 \div \frac{2}{3} = 3$.

To check the division, we multiply: $3 \cdot \frac{2}{3} = \frac{6}{3} = 2$. Since we got the original dividend, it checks.

We can use measurement division to check whether an answer to a division is reasonable.

For example, if I told you that $7 \div 1\frac{2}{3}$ equals $14\frac{1}{3}$, you can immediately see it doesn't make sense:

1 2/3 surely does not fit into 7 that many times. Maybe three to four times, but not 14!

You could also multiply to see that: 14-*and-something* times 1-*and-something* is way more than 14, and closer to 28 than to 14, instead of 7.

1. Find the answers that are unreasonable without actually dividing.

   a. $\frac{4}{5} \div 6 = \frac{2}{15}$ b. $2\frac{3}{4} \div \frac{1}{4} = \frac{7}{12}$ c. $\frac{7}{9} \div 2 = \frac{7}{18}$ d. $8 \div 2\frac{1}{3} = 18\frac{1}{3}$ e. $5\frac{1}{4} \div 6\frac{1}{2} = 3\frac{1}{8}$

2. Solve with the help of the visual model, checking how many times the given fraction fits into the other number. Then write a division. Lastly, write a multiplication that checks your division.

   a. How many times does  go into ?

   $2 \div \frac{3}{4} =$

   Check: ____ $\cdot \frac{3}{4} =$

   b. How many times does  go into ?

   ___ ÷ ___ =

   Check:

   c. How many times does 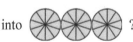 go into 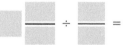 ?

   $3 \div$ ___ =

   Check:

   d. How many times does  go into  ?

   ___ ÷ ___ =

   Check:

3. Solve. Think how many times the fraction goes into the whole number. Can you find a *pattern* or a *shortcut*?

a. $3 \div \frac{1}{6} =$	b. $4 \div \frac{1}{5} =$	c. $3 \div \frac{1}{10} =$	d. $5 \div \frac{1}{10} =$
e. $7 \div \frac{1}{4} =$	f. $4 \div \frac{1}{8} =$	g. $4 \div \frac{1}{10} =$	h. $9 \div \frac{1}{8} =$

The shortcut is this:

$$5 \div \frac{1}{4} \quad\quad 3 \div \frac{1}{8} \quad\quad 9 \div \frac{1}{7}$$
$$\downarrow \downarrow \quad\quad\quad \downarrow \downarrow \quad\quad\quad \downarrow \downarrow$$
$$5 \cdot 4 = 20 \quad\quad 3 \cdot 8 = 24 \quad\quad 9 \cdot 7 = 63$$

Notice that 1/4 inverted (upside down) is 4/1 or simply 4. We call 1/4 and 4 reciprocal numbers, or just reciprocals. So the shortcut is: multiply by the reciprocal of the divisor.

Does the shortcut make sense to you? For example, consider the problem $5 \div (1/4)$. Since 1/4 goes into 1 exactly four times, it must go into 5 exactly $5 \cdot 4 = 20$ times.

Two numbers are reciprocal numbers (or reciprocals) of each other if, when multiplied, they make 1.

$\frac{3}{4}$ is a reciprocal of $\frac{4}{3}$, because $\frac{3}{4} \cdot \frac{4}{3} = \frac{12}{12} = 1$.	$\frac{1}{7}$ is a reciprocal of 7, because $\frac{1}{7} \cdot 7 = \frac{7}{7} = 1$.

You can find the reciprocal of a fraction $\frac{m}{n}$ by flipping the numerator and denominator: $\frac{n}{m}$.

This works, because $\frac{m}{n} \cdot \frac{n}{m} = \frac{n \cdot m}{m \cdot n} = \frac{m \cdot n}{m \cdot n} = 1$.

To find the reciprocal of a mixed number or a whole number, first <u>write it as a fraction</u>, then "flip" it.

Since $2\frac{3}{4} = \frac{11}{4}$, its reciprocal number is $\frac{4}{11}$. And since $28 = \frac{28}{1}$, its reciprocal number is $\frac{1}{28}$.

4. Find the reciprocal numbers. Then write a multiplication with the given number and its reciprocal.

a. $\frac{5}{8}$	b. $\frac{1}{9}$	c. $1\frac{7}{8}$	d. 32	e. $2\frac{1}{8}$
$\frac{5}{8} \cdot \underline{} = 1$	$\underline{} \cdot \underline{} = 1$	$\underline{} \cdot \underline{} = 1$	$32 \cdot \underline{} = 1$	$\underline{} \cdot \underline{} = 1$

5. Write a division sentence to match each multiplication above.

a. $1 \div \underline{} = \underline{}$	b. $1 \div \underline{} = \underline{}$	c. $1 \div \underline{} = \underline{}$	d. $\underline{} \div \underline{} = \underline{}$	e. $\underline{} \div \underline{} = \underline{}$

SHORTCUT: instead of dividing, multiply by the reciprocal of the divisor.

Study the examples to see how this works.

How many times does ⊘ go into ⊕?

$$\frac{3}{4} \div \frac{1}{3}$$
↓ ↓
$$\frac{3}{4} \cdot 3 = \frac{9}{4} = 2\frac{1}{4}$$

Answer: 2 1/4 times.

Does it make sense?

Yes, ⊘ fits into ⊕ a little more than two times.

How many times does 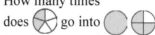 go into ?

$$\frac{7}{4} \div \frac{2}{5}$$
↓ ↓
$$\frac{7}{4} \cdot \frac{5}{2} = \frac{35}{8} = 4\frac{3}{8}$$

Answer: 4 3/8 times.

Does it make sense?

Yes. ⊗ goes into 1 3/4 over four times.

How many times does 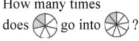 go into ?

$$\frac{2}{9} \div \frac{2}{7} =$$
↓ ↓
$$\frac{\cancel{2}}{9} \cdot \frac{7}{\cancel{2}} = \frac{7}{9}$$

Answer: 7/9 of a time.

Does it make sense?

Yes, because ⊛ does not go into ⊛ even one full time!

Remember: There are *two* changes in each calculation:

1. **Change the division into multiplication.**

2. **Use the reciprocal of the divisor.**

6. Solve these division problems using the shortcut. Remember to check to make sure your answer makes sense.

a. $\frac{3}{4} \div 5$

   ↓ ↓

   $\frac{3}{4} \cdot \frac{1}{5} =$

b. $\frac{2}{3} \div \frac{6}{7}$

c. $\frac{4}{7} \div \frac{3}{7}$

d. $\frac{2}{3} \div \frac{3}{5}$

e. $4 \div \frac{2}{5}$

f. $\frac{13}{3} \div \frac{1}{5}$

Now let's try to **make some sense visually** out of how reciprocal numbers fit into the division of fractions.

Example 1. We can think of the division **1 ÷ (2/5)** as asking, **"How many times does 2/5 fit into 1?"**

Using pictures: How many times does ◖ go into ⬤ ? (From the looks of it, at least two times!)

From the picture we can see that ◖ goes into ⬤ two times, and then we have 1/5 left over.

But how many times does $\frac{2}{5}$ fit into the leftover piece, $\frac{1}{5}$? How many times does ◖ go into △ ?

That is like trying to fit a TWO-part piece into a hole that holds just ONE part.
Only 1/2 of the two-part piece fits! So, 2/5 fits into 1/5 exactly half a time.

So we found that, in total, 2/5 fits into 1 exactly **2 ½ times**. We can write the division $1 \div \frac{2}{5} = 2\frac{1}{2}$ or $\frac{5}{2}$.
Notice, we got $1 \div \frac{2}{5} = \frac{5}{2}$. Checking that with multiplication, we get $\frac{5}{2} \cdot \frac{2}{5} = 1$. Reciprocals!

Example 2. We can think of the division **1 ÷ (5/7)** as, **"How many times does 5/7 fit into 1?"**

Using pictures: How many times does ◕ go into ⬤ ? (It looks like, a bit over one time.)

From the picture we can see that ◕ goes into ⬤ just once, and then we have 2/7 left over.

But how many times does $\frac{5}{7}$ fit into the leftover piece, $\frac{2}{7}$? How many times does ◕ go into ◗ ?

The five-part piece fits into a hole that is only big enough for two parts just 2/5 of the way.

So 5/7 fits into one exactly 1 2/5 times—and this makes sense because, as we noted at first, it looked like 5/7 fit into one a little over one time. The division is $1 \div \frac{5}{7} = 1\frac{2}{5}$ or $1 \div \frac{5}{7} = \frac{7}{5}$. Reciprocals again!

7. Write a division.

a. How many times does ◗ go into ⬤ ? $1 \div \frac{\square}{\square} =$ *Check:* Does your answer make sense visually?	**b.** How many times does ◹ go into ⬤ ? $1 \div \frac{\square}{\square} =$ *Check:* Does your answer make sense visually?
c. How many times does ▭ go into ▦ ? $1 \div \frac{\square}{\square} =$ *Check:* Does your answer make sense visually?	**d.** How many times does ⌐ go into ▦ ? $1 \div \frac{\square}{\square} =$ *Check:* Does your answer make sense visually?

8. Fill in the answers and complete the patterns. You will be able to do a lot of these in your head!

a.	b.	c.	d.
$3 \div \frac{1}{5} =$	$6 \div \frac{1}{4} =$	$1 \div \frac{1}{4} =$	$8 \div \frac{1}{2} =$
$3 \div \frac{2}{5} =$	$6 \div \frac{2}{4} =$	$2 \div \frac{1}{4} =$	$8 \div \frac{2}{2} =$
$3 \div \frac{3}{5} =$	$6 \div \frac{3}{4} =$	$3 \div \frac{1}{4} =$	$8 \div \frac{3}{2} =$
$3 \div \frac{4}{5} =$	$\square \div \frac{\square}{\square} =$	$\square \div \frac{\square}{\square} =$	$\square \div \frac{\square}{\square} =$
$3 \div \frac{5}{5} =$	$\square \div \frac{\square}{\square} =$	$\square \div \frac{\square}{\square} =$	$\square \div \frac{\square}{\square} =$

Epilogue (optional)

The lesson didn't go into full details as to why multiplication by the reciprocal always gives us the answer to a division problem. Let's continue that discussion a bit.

Any division can be turned into a multiplication. For example, from the division $2 \div \frac{3}{4} =$ _____ , we can write the multiplication $\frac{3}{4} \cdot$ _____ $= 2$.

To find what goes on the empty line, we can first of all put there the reciprocal of ¾: $\frac{3}{4} \times \frac{4}{3} \times$ _____ $= 2$.

Notice the multiplication of the two fractions above equals 1 (since they are reciprocals). To make the left side of the equation equal 2, we place 2 in the empty line: $\frac{3}{4} \times \frac{4}{3} \times 2 = 2$

And thus, the answer to the original division is $\frac{4}{3} \cdot 2$ (or $2 \cdot \frac{4}{3}$) — which is the original dividend times the reciprocal of the divisor.

Let's take another example: $2\frac{1}{6} \div 5\frac{3}{4} =$ _____

We turn it around and make a multiplication problem: $5\frac{3}{4} \cdot$ _____ $= 2\frac{1}{6}$.

Now, 5 ¾ = 23/4. So first, we insert the reciprocal of 23/4, which is 4/23: $5\frac{3}{4} \cdot \frac{4}{23} \cdot$ _____ $= 2\frac{1}{6}$.

Since 5 ¾ · (4/23) = 1, then the number we still need to put on the empty line must be 2 1/6, and thus the answer to the original division problem is 4/23 · (2 1/6), or (2 1/6) · 4/23 — the original divided times the reciprocal of the divisor.

We could use a similar argument to show in the general case that the answer to any division problem $a \div b$ is always going to be a times the reciprocal of b.

Divide Fractions

SHORTCUT: instead of dividing, multiply by the reciprocal of the divisor.

This shortcut works *in every case*, whether the numbers involved are whole numbers, fractions, or mixed numbers.

$\frac{2}{5} \div \frac{7}{9}$ $\downarrow \downarrow$ $\frac{2}{5} \cdot \frac{9}{7} = \frac{18}{35}$ Check: $\frac{18}{35} \cdot \frac{7}{9} = \frac{2}{5}$	$7 \div \frac{9}{10}$ $\downarrow \downarrow$ $7 \cdot \frac{10}{9} = \frac{70}{9} = 7\frac{7}{9}$ Check: $\frac{70}{9} \cdot \frac{9}{10} = \frac{7}{1} = 7$	$1\frac{10}{11} \div 5$ $\downarrow \downarrow$ $\frac{21}{11} \cdot \frac{1}{5} = \frac{21}{55}$ Check: $\frac{21}{55} \cdot 5 = \frac{21}{11} = 1\frac{10}{11}$

Notice: when you check the problems, you will need to use the *original* divisor, not the "flipped" one.

1. Solve. Change mixed numbers to fractions before dividing. Check each division by multiplication.

a. $\frac{9}{10} \div \frac{2}{5}$

Check:

b. $\frac{3}{7} \div \frac{4}{3}$

Check:

c. $\frac{2}{11} \div \frac{2}{3}$

Check:

d. $1\frac{7}{8} \div \frac{3}{4}$

Check:

e. $2\frac{1}{15} \div 1\frac{3}{5}$

Check:

f. $5\frac{10}{11} \div 6$

Check:

2. How many 2/3 cup servings can you get out of 5 cups of ice cream?

We have already looked at one meaning of division called **measurement division**, where we consider how many times this fits into that. For example, we can interpret 1 ÷ 1/5 as meaning how many copies of 1/5 fit into one.

Another meaning or interpretation of division is **equal sharing**, or "dividing equally among so many people." In this case, the divisor will be a *whole number*.

Example 1. Divide 8/10 of a pie among four people. The division is $\frac{8}{10} \div 4 = \frac{2}{10}$. Each person gets 2/10.

Example 2. Solve $\frac{6}{7} \div 4$. We have six slices (each slice being a seventh) and four people. Each person gets one slice, first of all, and then we have 2 slices left. We split those. So, each person gets 1 1/2 slices.

In fraction terms, the 1/2 slice is a fourteenth-part and the 1 slice becomes 2/14. Each person gets a total of 3/14 of the whole.

3. Solve these problems by *reasoning logically*. Write a division sentence for each problem.

a. There is 1 4/6 of a pizza left over and two people share it equally. How much does each one get?	**b.** There is 9/10 of a cake left over and three people share it equally. How much does each one get?

4. The picture shows how much pie is left. That amount is divided among a certain number of people. How much does each person get? Write a division sentence.

a. Divide among three people:	**b.** Share among three people:	**c.** Divide among six people:
d. Divide between two people:	**e.** Divide among five people:	**f.** Share among four people:

5. Three people share a 1/4-kg chocolate bar equally. How many kilograms will each of them get?

6. Five siblings inherited a plot of land that measures 2 4/10 acres. If they divide the plot equally, what portion of an acre will each one get?

The following problems may require division or multiplication of fractions. Think carefully.

7. Among many other ingredients, a recipe calls for 2/3 cup of wheat flour for each full recipe. In her pantry, Sarah had plenty of all the other ingredients, but only a small portion of wheat flour. How many batches of the recipe can she make if she has...

   a. 1/3 cup of wheat flour?

   b. 1 cup of wheat flour?

8. An airport takes up a rectangular area that is 2 1/8 miles long and 1/2 mile wide. What is its area?

9. **a.** Which calculation on the right can be used to solve the problem?

   Jeannie likes to eat 2/5 of a package of crackers in one sitting. How many times can she do that, if she has three packages of crackers in her cupboard?

   (i) $\frac{2}{5} \div 3$ (ii) $\frac{2}{5} \cdot 3$ (iii) $3 \div \frac{2}{5}$

   b. Notice that the answer to the question needs to be *whole number*, yet the calculation may not give you that. Do the calculation. What does a fractional result mean in this context?
   (Hint: a drawing can help you interpret the result.)

10. A gym class takes 3/4 of an hour.

    a. How many such classes could you fit, end-to-end, in five hours?

    b. If those classes start at 9 AM, when will the last class end?

11. Mary's vegetable garden is 2 ½ meters by 2 ½ meters. She divided her garden into quarters in order to plant four different vegetables. Look at the different calculations for finding the area of each of those quarters.

(i)	(ii)	(iii)
Total area: $\frac{5}{2}$ m · $\frac{5}{2}$ m = $\frac{25}{4}$ m² **Area of each quarter:** $\frac{25}{4}$ m² ÷ 4 = $\frac{25}{16}$ m² = 1 $\frac{9}{16}$ m²	**Area of each quarter:** 250 cm · 250 cm ÷ 4 = 15,625 cm²	**Area of each quarter:** 2.5 m · 2.5 m ÷ 4 = 1.5625 m²

   a. Which calculation is correct (or are several)?

   b. Which one is the most useful or natural? Why?

12. An airport runway is two miles long, and takes up 1/16 square mile in area. How wide is it, in miles?

    In feet?

13. The sides of a rectangle are in a ratio of 2:3, and its perimeter is 1 1/4 inches.

    a. What are the lengths of its sides?

    b. Draw the rectangle.

    Hint: draw a sketch.

Puzzle Corner

How many 3/8-meter long pieces can you cut out of 11 meters of ribbon?

How long is the piece that is left over?

Problem Solving with Fractions 1

1. **a.** Anna needs to make 30 servings of spiced coffee for a party. Calculate the amount of each ingredient she needs.

Spiced Coffee – 4 servings
1 1/2 teaspoons of ground cinnamon
1/2 of a teaspoon of ground nutmeg
2 tablespoons of sugar
1 cup of heavy cream
3 cups of coffee
4 teaspoons of chocolate syrup

   b. Next week she wants to make just *one* serving for herself. Calculate the amount of each ingredient she needs.

2. Sam planted tomatoes in his garden, which is a rectangle with an area of 2 ½ m². If one side of the garden measures 5 m, how long is the other side?

3. Which is a better deal: a book that costs $45.55 at 1/5 off or a book that costs $52.80 at 1/4 off?

Example. Richard paid 3/10 of his $1,140 paycheck in taxes. Of what remained, he paid 1/6 on a loan payment. How much did he have left after those payments?

We will solve this in two steps:

1. First find out how much Richard has left after paying taxes.
2. Then find out how much he has left after paying the loan payment.

1. There are several different ways to find out how much Richard has left after paying taxes:

  (i) We could calculate what 3/10 of $1,140 is, and subtract that from $1 140. First, 1/10 of $1,140 is $114. Then, 3/10 is three times as much, or $342. Lastly, subtract $1,140 − $342 = $798.

  (ii) Since taking away 3/10 of his paycheck leaves 7/10 of it, we could just calculate what 7/10 of $1,140 is. So 1/10 of $1,140 is $114, and 7 · $114 is $798.

  (iii) We could use decimals and calculate 0.7 · $1,140 = $798.

  So Richard has $798 left after taxes.

2. His loan payment is 1/6 of $798. We can easily calculate 1/6 of $798 by dividing: $798 ÷ 6 = $133. Subtract that from $798, and we find that Richard has $798 − $133 = $665 left.

4. The unknown is given as a part of a part. Solve for *x*.

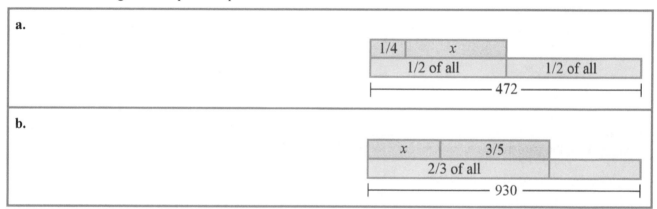

Draw a bar model to help you solve this problem.

5. Elaine gave her grandmother 3/4 of the 48 flowers she had picked.
   Then her grandmother gave 1/4 of *those* flowers to her neighbor.
   How many flowers does Elaine's grandmother have now?
   How many flowers did the neighbor get?
   How many flowers does Elaine have left?

Draw a bar model to help you solve these problems.

6. Father paid 1/5 of his paycheck in taxes. After that, he used 1/6 of what remained as a loan payment. Then he had $860 left. How much was his paycheck?

7. Father used 1/6 of his $1,200 to pay for car repairs. Of what was left, he used 2/5 to buy groceries. How much money does Father have left now?

8. **a.** Of the total retail sales of potatoes, the farmer gets 1/8, the wholesale dealer gets 1/12 and the store owner gets the rest. How much (as a fraction) does the store owner get?

   b. If the total sales were $4,500, how much (in dollars) would the farmer, the wholesale dealer and the store owner each get?

Problem Solving with Fractions 2

Example 1. The two sides of a rectangle are in a ratio of 2:3. The perimeter is 7 1/2 inches. How long are the sides?

Let's draw a sketch of the rectangle and mark the sides as 2 and 3 "parts" long. We can see that the perimeter is 2 + 3 + 2 + 3 = 10 parts long. So, each part is 7 1/2 in ÷ 10. We can find the length of one part by dividing fractions: $7\frac{1}{2} \div 10 = \frac{15}{2} \cdot \frac{1}{10} = \frac{3}{4}$.

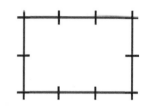

Remember that 3/4 inch is *not* our final answer! That is just the length of one "part." The short side is 2 parts, so it is 2 × 3/4 in, and the long side is 3 parts, so it is 3 × 3/4 in. Therefore, the sides are 1 1/2 in and 2 1/4 in long.

Check the length of the perimeter: 1 1/2 in + 2 1/4 in + 1 1/2 in + 2 1/4 in = 7 1/2 in, as given.

1. Emily mixed 1 part concentrate with 7 parts water to make a total of 52 ounces of juice. How many ounces of concentrate and how many ounces of water were in the juice?

2. A rectangle has a perimeter of 10 ½ units, and its aspect ratio (the ratio of its width to its length) is 1:5. Find the lengths of the two different sides.

3. Heather has two different routes that she can walk to the local swimming pool, and their lengths are in a ratio of 3:4. The shorter route is 1 1/8 miles long. Find the length of the longer route.

Example 2. Two-sevenths of a class stayed home, so only 25 students showed up at school. How many students are in the whole class?

If 2/7 of the class stayed home, then 5/7 of the class (25 students) came to school. So 25 is 5/7 of the whole group. See the bar model.

We can now solve one "part" (which is 1/7 of the class) by the division $25 \div 5 = 5$.
Therefore, 7/7 of the class—or the whole class—is $7 \cdot 5 = 35$ students.

4. Fill in the blanks.

a. 3/4 of a number is 15.	**b.** 2/9 of a number is 24.	**c.** 7/8 of a number is 49.
1/4 of that number is _____.	1/9 of that number is _____.	1/8 of that number is _____.
The number is _____.	The number is _____.	The number is _____.

5. Use reasoning similar to what you used in the previous exercise. In each case, find the number.

a. 4/5 of a number is 24.	**b.** 2/3 of a number is 40.
c. 8/11 of a number is 56.	**d.** 5/7 of a number is 45.

6. Of a horse club's members, 7/9 are girls, and 8 (the rest) are boys. How many members does the club have?

7. It took 21 gallons of gas for a tractor to plow 3/5 of a farm. How much gas will be needed to plow the rest of the farm?

8. One-third of the audience at a concert were seniors who had half-price tickets. The total audience count was 657, and a full-price ticket cost $24.50. Find the total income from ticket sales for the concert.
   Hint: Start out by finding how many were seniors and how many were not.

9. Try to visualize this problem by drawing or by using physical objects.

   a. An eraser is 1/8 inch thick. How many such erasers can be stacked into a 4-inch tall box?

   b. The eraser is 1 3/8 inches long. The box is 6 inches long. How many erasers fit in lengthwise?

   c. The eraser is 13/16 inch wide. The box is 5 inches wide. How many erasers fit side by side in the width of the box?

   d. Use the calculations above to figure out what would be the total number of erasers that could fit into the box.

10. How many of the same kind of erasers would fit into a box whose dimensions are 12 in (length) by 10 in (width) by 8 in (height)?

11. First, a length of fabric was cut into two equal halves. Then 3/4 of one of the halves was cut off, leaving a 17-centimeter piece. How long was the fabric originally?

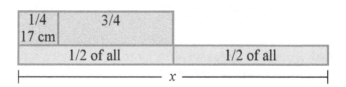

12. Jeremy gave 1/3 of his apple harvest to Dave, who gave half of his share to a neighbor. The neighbor got 15.5 kg. How many kilograms of apples did Jeremy harvest?

Chapter 7 Mixed Review

1. Jane mixed 2 parts of concentrated juice with 6 parts of water to make a total of 64 ounces of juice. How many ounces of concentrate and how many ounces of water were in the juice?
   (Ratio Problems and Bar Models 1/Ch.4)

2. Write the equivalent rates. (Using Equivalent Rates/Ch.4)

a. $\dfrac{\$80}{4\text{ hr}} = \dfrac{}{1\text{ hr}} = \dfrac{}{3\text{ hr}} = \dfrac{}{15\text{ min}}$	b. $\dfrac{2\text{ m}^2}{5\text{ min}} = \dfrac{10\text{ m}^2}{} = \dfrac{}{5\text{ hours}} = \dfrac{250\text{ m}^2}{}$

3. A mixture of salt and water weighs 1.2 kg. It contains 2% salt by weight, and the rest is water. How many grams of salt and how many grams of water are in the mixture?
   (Percentage of a Number Using Decimals/Ch.5)

4. A train traveled 165 miles from one town to the next at an average speed of 90 mph. When did the train leave, if it arrived at 1440 hours (2:40 P.M.)?
   (Using Equivalent Rates/Ch.4)

5. Multiply or divide the decimals by the powers of ten. (Multiply and Divide by Powers of Ten/Ch.3)

a.	b.	c.
$10 \cdot 0.3909 =$	$1.08 \cdot 100 =$	$10^6 \cdot 8.02 =$
$1{,}000 \cdot 4.507 =$	$0.0034 \cdot 10^4 =$	$10^5 \cdot 0.004726 =$
d.	e.	f.
$0.93 \div 100 =$	$3.04 \div 1{,}000 =$	$98.203 \div 10^5 =$
$48 \div 10 =$	$450 \div 10^4 =$	$493.2 \div 10^6 =$

6. Factor the following composite numbers into their prime factors.
(The Sieve of Eratosthenes and Prime Factorization/Ch.6)

a. 65 /\	b. 75 /\	c. 82 /\

7. Find a number between 640 and 660 that is divisible by 3 and 7.
(The Sieve of Eratosthenes and Prime Factorization/Ch.6)

8. First, find the GCF of the numbers. Then factor the expressions using the GCF. (Factoring Sums/Ch.6)

a. GCF of 16 and 42 is _____ 16 + 42 = ____ (____ + ____)	b. GCF of 98 and 35 is _____ 98 + 35 = ____ (____ + ____)

9. Draw two rectangles, side by side, to represent the sum 18 + 30. (Factoring Sums/Ch.6)

10. Calculate the values of y according to the equation $y = 2x - 5$. (Using Two Variables/Ch.2)

x	3	4	5	6	7	8
y						

Now, plot the points.

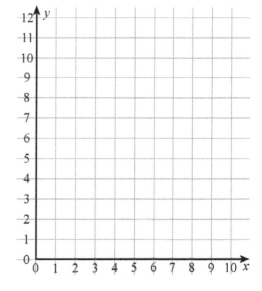

Fractions Review

1. Add.

a. $\dfrac{5}{12} + \dfrac{1}{3}$	b. $\dfrac{5}{7} + \dfrac{1}{6}$	c. $1\dfrac{3}{5} + \dfrac{7}{8}$

2. Subtract. First write equivalent fractions with the same denominator.

a. $6\dfrac{2}{3} \rightarrow$ $- 2\dfrac{1}{6} \rightarrow -$	b. $7\dfrac{1}{6} \rightarrow$ $- 2\dfrac{3}{5} \rightarrow -$	c. $8\dfrac{9}{11} \rightarrow$ $- 4\dfrac{1}{3} \rightarrow -$

3. The pictures show how much pizza is left. Find the given part of it. Write a multiplication sentence.

a. Find $\dfrac{3}{4}$ of	b. Find $\dfrac{1}{5}$ of	c. Find $\dfrac{2}{3}$ of

4. Multiply. Shade the areas to illustrate the multiplication.

a.  $\dfrac{1}{4} \cdot \dfrac{3}{4} =$	b.  $\dfrac{3}{4} \cdot \dfrac{6}{7} =$

5. Simplify before you multiply.

a. $\dfrac{9}{12} \cdot \dfrac{6}{15}$ b. $\dfrac{3}{20} \cdot \dfrac{4}{21}$ c. $\dfrac{14}{40} \cdot \dfrac{10}{42}$

6. Write a division sentence for each problem and solve.

a. How many times does  go into ?	b. How many times does go into ?

7. Fill in the blanks and give an example. You can choose *any* number to divide by 4.

Dividing a number by 4 is the same as multiplying it by ____. Example:

8. Solve.

a. $\dfrac{2}{3} \div \dfrac{1}{5}$	b. $2\dfrac{1}{7} \div 1\dfrac{1}{2}$	c. $6 \div 1\dfrac{2}{3}$

9. A small, rectangular garden plot measures 7 1/2 feet by 4 3/8 feet.

   a. Find its area.

   b. Find its perimeter.

10. Write a real-life situation to match this fraction division: $\dfrac{9}{12} \div 3 = \dfrac{3}{12}$

11. How many ¾ pound packages of ground beef can you make out of 10 pounds of ground beef?

12. Two-and-a-half gallons of ice cream is divided (unequally) into two parts that are in the ratio of 1:7. How much is each part?

13. Five-sixths of the class went outside for recess, and 6 students stayed in the classroom. How many students are in the whole class?

14. Two-fifths of a certain number is 160. What is the number?

15. Two farmers divided a day's kiwi fruit harvest. One farmer got 2/5 of the harvest and the other farmer got the rest. The farmer who got the least gave 1/3 of his kiwi to his son, and kept 12 kilograms. How many kilograms was the day's kiwi fruit harvest?

Puzzle Corner

a. Solve this "long" division!

$$\frac{1}{2} \div 5 \div 4 \div 3 \div 2 =$$

b. What did this division start with?

$$\underline{} \div 3 \div 5 \div 7 \div 9 = \frac{1}{1,260}$$

Chapter 8: Integers
Introduction

In chapter 8, students are introduced to integers, the coordinate plane in all four quadrants and integer addition and subtraction. The multiplication and division of integers will be studied next year.

Integers are introduced using the number line to relate them to the concepts of temperature, elevation and money. We also study briefly the ideas of absolute value (an integer's distance from zero) and the opposite of a number.

Next, students learn to locate points in all four quadrants and how the coordinates of a figure change when it is reflected across the x or y-axis. Students also move points according to given instructions and find distances between points with the same first coordinate or the same second coordinate.

Adding and subtracting integers is presented through two main models: (1) movements along the number line and (2) positive and negative counters. With the help of these models, students should not only learn the shortcuts, or "rules", for adding and subtracting integers, but also understand *why* these shortcuts work.

A lesson about subtracting integers explains the shortcut for subtracting a negative integer from three different viewpoints (as a manipulation of counters, as movements on a number-line and as a distance or difference). There is also a roundup lesson for addition and subtraction of integers.

The last topic in this chapter is graphing. Students will plot points on the coordinate grid according to a given equation in two variables (such as $y = x + 2$), this time using also negative numbers. They will notice the patterns in the coordinates of the points and the pattern in the points drawn in the grid and also work through some real-life problems.

You will find free videos covering many topics of this chapter at **https://www.mathmammoth.com/videos/** (choose 6th grade).

The Lessons in Chapter 8

	page	span
Integers	73	*3 pages*
Coordinate Grid	76	*4 pages*
Coordinate Grid Practice	80	*3 pages*
Addition and Subtraction as Movements	83	*3 pages*
Adding Integers: Counters	86	*3 pages*
Subtracting a Negative Integer	89	*2 pages*
Add and Subtract Roundup	91	*2 pages*
Graphing	93	*4 pages*
Chapter 8 Mixed Review	97	*2 pages*
Integers Review	99	*3 pages*

Helpful Resources on the Internet

We have compiled a list of Internet resources that match the topics in this chapter. This list of links includes web pages that offer:

- **online practice** for concepts;
- online **games**, or occasionally, printable games;
- **animations** and interactive **illustrations** of math concepts;
- **articles** that teach a math concept.

We heartily recommend you take a look at the list. Many of our customers love using these resources to supplement the bookwork. You can use the resources as you see fit for extra practice, to illustrate a concept better, and even just for some fun. Enjoy!

https://l.mathmammoth.com/gr6ch8

Integers

When we continue the number-line towards the left from zero, we come to the **negative numbers**.

The **negative whole numbers** are −1, −2, −3, −4 and so on.
The **positive whole numbers** are 1, 2, 3, 4 and so on. You can also write them as +1, +2, +3, *etc*.
Zero is neither positive nor negative.
All of the negative and positive whole numbers and zero are called **integers**.

Read −1 as "negative one" and −5 as "negative five". Some people read −5 as "minus five".
That is very common, and it is not wrong, but be sure that you do not confuse it with subtraction.

Put a "−" sign in front of negative numbers. This sign can also be elevated: ⁻5 is the same as −5.

Often, we need to put brackets around negative numbers in order to avoid confusion with other symbols.
Therefore, ⁻5, −5 and (−5) all mean "negative five".

Negative numbers are commonly used with temperature. They are also used to express debt. If you owe $5, you write that as −$5. Another use is with elevation below sea level. For example, just as 200 m can mean an elevation of 200 meters above sea level, −100 m would mean 100 meters *below* sea level.

1. Plot the integers on the number-line.   **a.** −7   **b.** +6   **c.** −4   **d.** −2

2. Write an integer appropriate to each situation.

   a. Daniel owes $23.

   b. Mary earned $250.

   c. The airplane flew at the altitude of 8 800 meters.

   d. The temperature in the freezer is 18 degrees Celsius below zero.

   e. A dolphin dove 9 m below sea level.

3. The temperature changed from what it was before. Find the new temperature.
   You can draw the mercury on the thermometer to help you.

before	1°C	2°C	−2°C	−4°C	−12°C	−8°C
change	drops 3°C	drops 7°C	drops 1°C	rises 5°C	rises 4°C	rises 3°C
now						

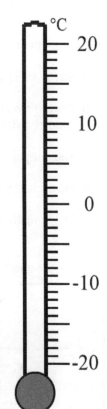

73

Which is more, −5 or −2?

Which is *warmer*, −5°C or −2°C? Clearly −2°C is. Temperatures just get colder and colder the more you move towards the negative numbers. We can write a comparison: −2°C > −5°C.

Which is the *better* money situation, to have −$5 (owe $5) or to have −$2 (owe $2)? Clearly, it is better to owe only $2 because you can pay that off easier. We can write: −$5 < −$2.

Which is the *higher* elevation, −5 m or −2 m? Of course, 2 m below sea level, or −2m, is higher.

On the number line, the number that is *farther to the right* is the **greater** number. So, −5 < −2.

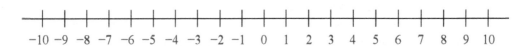

4. Compare. Write < or > between the numbers. You can plot the integers on the number line to help you.

a. −2 ☐ −3	b. 8 ☐ −8	c. −3 ☐ 0	d. 4 ☐ −3	e. −5 ☐ −9
f. −10 ☐ −30	g. −4 ☐ 1	h. 0 ☐ −13	i. −2 ☐ −7	j. −11 ☐ −14

5. You can use the number line to help you. Which integer is ...

   a. 2 more than −4   **b.** 5 more than −3   **c.** 3 less than 1   **d.** 6 more than −11

6. Find the number that is 5 less than ...   **a.** 0   **b.** −3   **c.** 3

7. Express the situations using integers. Then write > or < to compare them.

   a. Shelly owes $10 and Mary owes $8.

   b. One fish was swimming 3 m below the surface of the water, and another fish was swimming 4 m below the surface of the water.

   c. The temperature this morning was 10°C below zero. Now it is 6°C below zero.

   d. Henry has $5. Emma owes $5.

   e. The temperature during the day was 10°C but at night it was 2°C below zero.

8. Write the numbers in order from the least to the greatest.

a. −2 0 −4 4	b. −3 −6 5 3
c. −20 −10 −14 −9	d. −3 0 −6 −8

The **absolute value** of a number is its distance from zero.

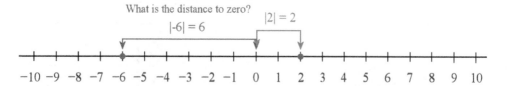

We denote the absolute value of a number using vertical bars around the number.

So, $|-4|$ means "the absolute value of 4", which is 4. Similarly, $|87| = 87$.

9. Find the absolute values of these numbers.

   a. $|-5|$ **b.** $|-12|$ **c.** $|7|$ **d.** $|0|$ **e.** $|68|$

The **opposite** of a number is the number that is at the same distance from zero as it is, but on the *opposite* side of the number-line (in regards to zero).

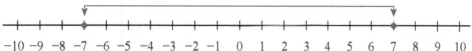

We denote the opposite of a number using the minus sign. For example, -4 means the opposite of 4, which is -4. Or, $-(-2)$ means the opposite of negative two, which is 2.

The opposite of zero is zero itself. In symbols, $-0 = 0$.

"But wait," you might ask, "doesn't -4 mean negative four, not the 'opposite of four'?"

It can mean either. Sometimes the context will help you to differentiate between the two (to tell which is which). Other times it's unnecessary to differentiate because, after all, the opposite of four is negative four: $-4 = -4$. ☺

So there are three different meanings for the minus sign:

1. To indicate subtraction: $7 - 2 = 5$.
2. To indicate negative numbers: "negative 7" is written -7.
3. To indicate the opposite of a number: $-(-14)$ is the opposite of negative 14.

10. Think of the minus sign as signifying "the opposite of". Simplify.

    a. -5 **b.** $-(-9)$ **c.** -10 **d.** -0 **e.** $-(-100)$

11. Write using mathematical symbols, and simplify (solve) if possible.

    a. the opposite of 6

    b. the opposite of the absolute value of 6

    c. the absolute value of negative 6

    d. the absolute value of the opposite of 6

    e. the opposite of -6.

    f. the absolute value of the opposite of -6

Coordinate Grid

This is the *coordinate grid* or *coordinate plane*. We have extended the *x*-axis and the *y*-axis to include negative numbers now. The axes cross each other at the *origin*, or the point (0, 0).

The axes divide the coordinate plane into four parts, called *quadrants*. Previously you have worked in only the so-called first quadrant, but now we will use all four quadrants.

The coordinates of a point are found in the same manner as before. Draw a vertical line (either up or down) from the point towards the *x*-axis. Where this line crosses the *x*-axis tells you the point's *x*-coordinate.

Similarly, draw a horizontal line (either right or left) from the point towards the *y*-axis. Where this line crosses the y-axis tells you the point's *y*-coordinate.

We list first the point's *x*-coordinate and then the *y*-coordinate. Look at the examples in the picture.

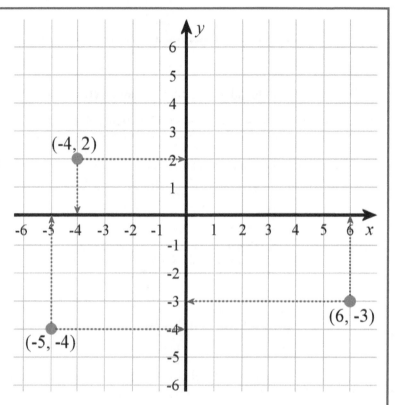

1. Write the *x*- and *y*-coordinates of the points.

   A (___ , ___)

   B (___ , ___)

   C (___ , ___)

   D (___ , ___)

   E (___ , ___)

   F (___ , ___)

   G (___ , ___)

   Self-check: Add the *x*-coordinates of all points. You should get −7.

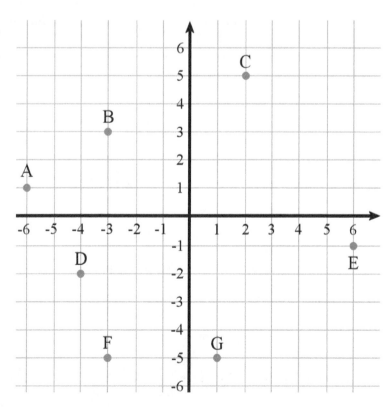

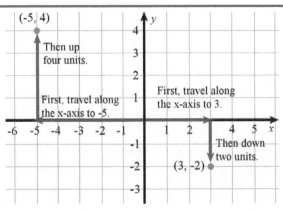
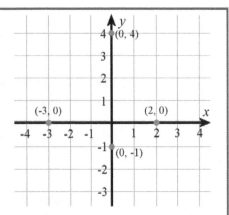

How to plot a point

1. First travel along the *x*-axis to the number of the *x*-coordinate.
2. Travel *up or down* the number of units of the *y*-coordinate.

   Naturally, if the *y*-coordinate is positive, you travel up. If it is negative, you travel down.

Remember:

If the point is <u>on</u> the *x*-axis, then its *y*-coordinate is zero.

If the point is <u>on</u> the *y*-axis, then its *x*-coordinate is zero.

2. Plot the following sets of points. Connect them with line segments to form figures. Which figures are formed?

   a. (−2, 0), (−2, 4), (2, 4), (2, 0)

   b. (−6, −6), (−5, −5), (−2, −5), (−3, −6)

   c. (2, −5), (3, −2), (5, −2), (8, −5)

   d. (−6, 0), (−3, 4), (0, 0)

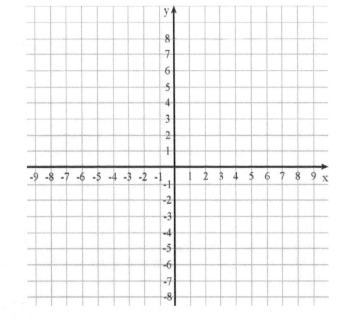

3. Find the area of the rectangle whose vertices are at (−1, 6), (−1, −2), (3, −2) and (3, 6).

4. Two vertices of a rectangle are (−5, −3) and (−2, 4). What are the other two vertices?

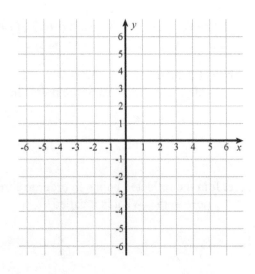

77

5. **a.** Plot the following points: (−6, 5), (−5, 3), (−3, 3), (−1, 4), (−2, 5). Then connect the points with line segments in the order they are given.

   b. For each point above, **change the *x*-coordinate into its opposite.** That means that −5 changes to 5, −1 changes to 1, and so on. Don't change the *y*-coordinates. Write the points here:

   c. Plot these new points. Connect them in the same order. What do you notice?

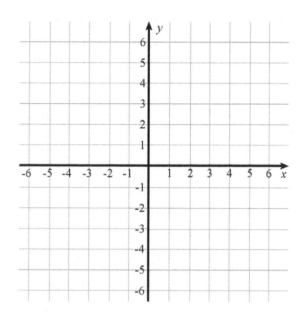

6. For each of these points (−6, 5), (−5, 3), (−3, 3), (−1, 4), (−2, 5), **change the *y*-coordinate into its opposite.**

   Plot these new points. Connect them in the same order. What do you notice?

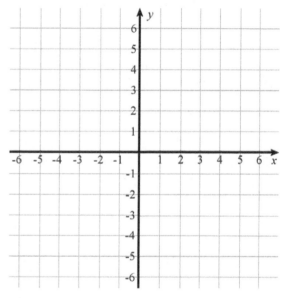

When we change the *x*-coordinate of a point into its opposite, without changing the *y*-coordinate (as in problem 5), the point gets *reflected* or *mirrored* in the *y*-axis.

When we change the *y*-coordinate of a point into its opposite, without changing the *x*-coordinate (as in problem 6), the point gets *reflected* or *mirrored* in the *x*-axis.

7. **a.** The vertices of a shape are (−6, 2), (−4, 6), (4, 6) and (−2, 2). Draw the shape.

   b. Reflect the points in the *x*-axis. Write the coordinates of the points here.

   c. Join the reflected points in order to form a new shape. What is it called?

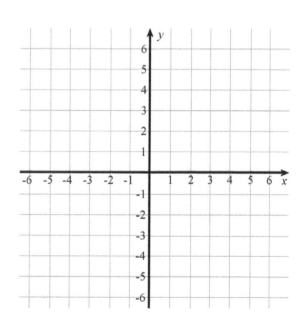

8. Reflect these points in the x-axis. What are the coordinates of the reflected points?

   a. (2, 7)　　　　b. (−15, 20)　　　　c. (−11, −21)　　　　d. (34, −19)

9. Reflect these points in the y-axis. What are the coordinates of the reflected points?

   a. (3, 9)　　　　b. (22, −20)　　　　c. (−67, −35)　　　　d. (−51, 60)

10. a. Draw a trapezoid with vertices of (4, 2), (7, 2), (8, 6) and (5, 6).

    b. Now mirror it in the y-axis.

    c. Reflect the same, original trapezoid in the x-axis.

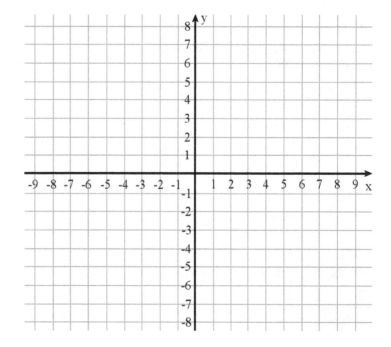

11. The points (−2, 1), (−8, 2), (−6, 3), (−8, 4) and (−2, 5) are the vertices of a pennant.

    a. Draw the pennant.

    b. Reflect it in the y-axis.

    c. After reflecting it, move the pennant five units down.

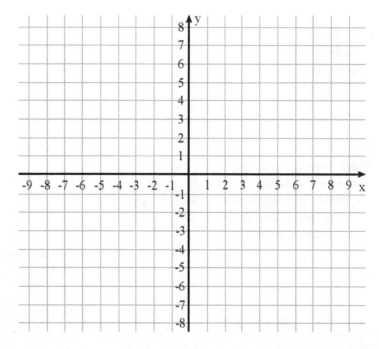

Puzzle Corner

Anne drew a secret figure. She moved it 2 units up, and then she mirrored it in the x-axis. Its vertices are now at: (−7, −7), (−4, −3) and (−2, −6). What were the coordinates of the original vertices?

Coordinate Grid Practice

Notice in the grid, the point (−6, 5) moves four units to the right. It ends up at (−2, 5).

1. **a.** The points (−5, −2), (−1, −7) and (1, −6) are vertices of a triangle. Draw the triangle.

    b. Move the triangle five units up (draw the new triangle). Write the coordinates of the moved vertices.

    (−5, −2) → (_____ , _____)

    (−1, −7) → (_____ , _____)

    (1, −6) → (_____ , _____)

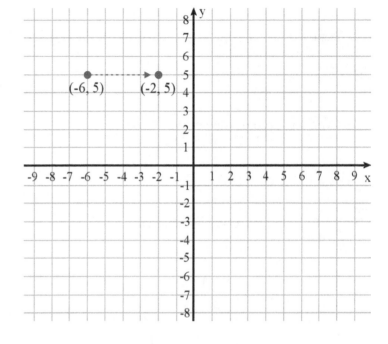

2. Write the coordinates of the new points based on the directions in the box on the right.

Point	Direction	New point
(1, 1)	7 units down	
(2, −2)	6 units left	
(−2, 7)	5 units right	
(−2, −2)	4 units down	

3. The point (−5, 5) is moved 8 units to the right *and* 3 units down. What are its new coordinates?

4. Jay drew a secret figure, and then he moved it 8 units up. The vertices of the moved figure are now at: (−4, 8), (−6, 6), (−4, 2) and (1, 6). What were the coordinates of the original vertices?

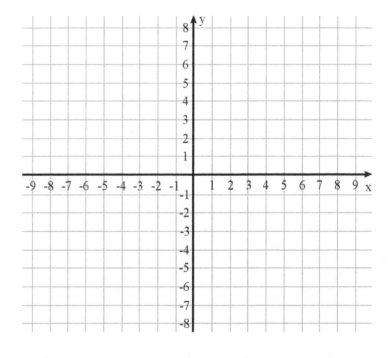

80

5. Find the difference between the two temperatures. You can use the thermometer to help you.

   a. −3°C and 3°C

   b. 4°C and −6°C

   c. −15°C and −7°C

   d. −4°C and 12°C

   e. −7°C and −28°C

   f. −3°C and 0°C

   g. 6°C and −6°C

   h. 0°C and −13°C

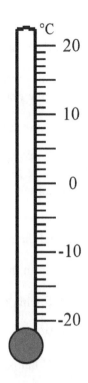

6. *Explain* in your own words how to find the **distance** between the numbers −29 and 28 on a number line.

7. Change your explanation above (if need be) so that you use *absolute value* in your explanation.

8. *Explain* in your own words how to find the distance between two negative numbers, such as −49 and −72, on the number line.

> To find the distance between two numbers where one is negative and the other is positive, add their absolute values. Remember, the absolute value tells us how far the number is from zero.
>
> **Example 1.** The distance between −99 and 99 is $|-99| + |99|$ or $99 + 99 = 198$.

> To find the distance between two negative numbers, you can simply find the distance between their opposites, which are both positive numbers.
>
> **Example 2.** The distance between −1 200 and −500 is the same as the distance between 1 200 and 500, which is 700 (you can subtract 1 200 − 500 to find that).

9. Find the distances between the points.

   a. (12, 56) and (12, −15)

   b. (−34, 9) and (−8, 9)

10. Find the perimeter of a rectangle with vertices (46, 50), (−22, 50), (−22, −17) and (46, −17).

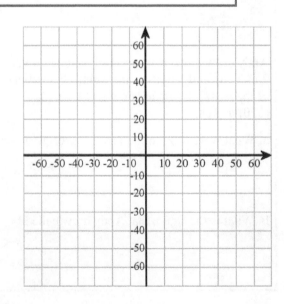

11. Lily drew a map of her neighborhood. She put her house at the origin, or the point (0, 0). Each square on the map is **50 m**.

    What are the coordinates of...
    (Remember that the gridlines are 50 m apart.)

    a. ...the school?

    b. ...Grace's house?

    c. ...the park?

    Find the distances:

    d. from the park to the service station.

    e. Sophia's house to Grace's house.

    f. What is the distance from (−200, 100) to (−25, 100)?

    g. What is the distance from (−300, 250) to (−300, −350)?

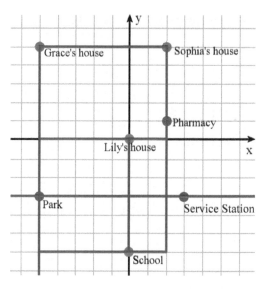

12. Activity (optional). Use a street map of any town where the streets are in a grid. Print it out and draw a coordinate grid with all four quadrants over it. Now, make a treasure hunt or a similar activity for your friend, giving locations using coordinates or using directions (go such-and-such a distance left, right, up, down, or north, south, east, west).

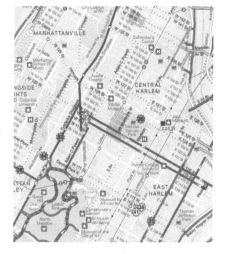

Puzzle Corner

Andy drew a secret figure, and then he moved it 2 units up and 7 units to the right. The vertices of the *moved* figure are now at: (0, 0), (3, 5), (5, 0) and (8, 5).

a. What were the coordinates of the original vertices?

b. What is the figure that Andy drew called?

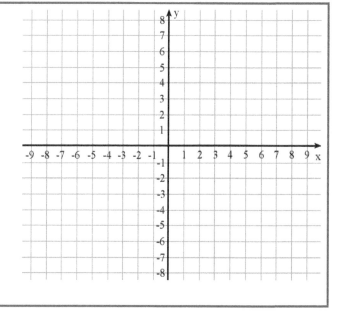

Addition and Subtraction as Movements

Suppose you are at 4. You jump 5 steps *to the right*. You end up at 9. We write *an addition*: 4 + 5 = 9.

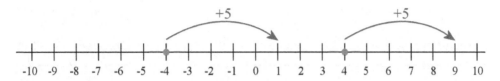

Now you are at −4. You jump 5 steps *to the right*. You end up at 1. We write *an addition*: −4 + 5 = 1.

Addition can be shown on the number line as a movement to the *right*.

1. Write an addition sentence (an equation) to match each of the number line jumps.

   a.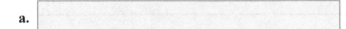
   b.

   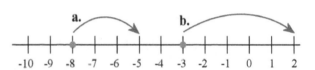

   c.
   d.

   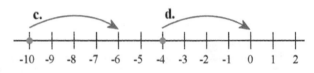

2. Draw a number line jump for each addition sentence.

   a. −8 + 2 = _____   **b.** −5 + 4 = _____

   c. −7 + 5 = _____   **d.** −10 + 12 = _____

3. Write an addition sentence.   Addition sentence:

   a. You are at −3. You jump 6 to the right. You end up at _____.

   b. You are at −8. You jump 8 to the right. You end up at _____.

   c. You are at −4. You jump 7 to the right. You end up at _____.

   d. You are at −10. You jump 3 to the right. You end up at _____.

You are at 4. You jump 5 steps *to the left*. You end up at −1. We write <u>a subtraction</u>: 4 − 5 = −1.

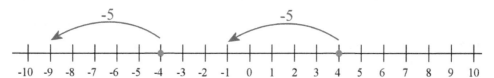

You are at −4. You jump 5 steps *to the left*. You end up at −9. We write <u>a subtraction</u>: −4 − 5 = −9.

Subtraction can be shown on the number line as a movement to the *left*.

Note: These three mean the same:

−4 − 5 = −9

(−4) − 5 = (−9)

⁻4 − 5 = ⁻9

We can use parentheses around a negative number if we need to make clear that the minus sign is for "negative", and not for subtraction. We can also use an elevated minus sign for clarity. However, in the above situation, there is no confusion, so the parentheses are not necessary and are usually omitted.

4. Write a subtraction sentence to match the number line jumps.

a.

b.

c.

d.

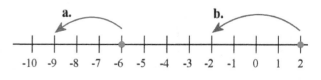

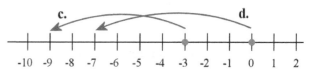

5. Draw a number line jump for each subtraction.

a. 1 − 5 = _____   **b.** 0 − 8 = _____

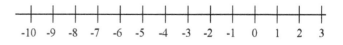

c. ⁻2 − 4 = _____   **d.** ⁻7 − 3 = _____

6. Write a subtraction sentence.   Subtraction sentence:

   a. You are at ⁻3. You jump 5 to the left. You end up at _____.

   b. You are at 5. You jump 10 to the left. You end up at _____.

   c. You are at ⁻5. You jump 5 to the left. You end up at _____.

84

Number line jumps with mixed addition and subtraction

- The first number tells you where you *start*.
- Next comes the sign: a *plus* sign tells you to jump *right*, and a *minus* sign tells you to jump *left*.
- Next comes the number of *steps* to jump.

Notice that the number of steps that you jump is *not* negative.

| ⁻2 + 6 means:
Start at ⁻2 and move 6 steps to the right.
You end up at 4.

You started out negative, but you moved towards the positives, and you ended up on the positive side! | ⁻2 − 6 means:
Start at ⁻2 and move 6 steps to the left.
You end up at −8.

You started out negative at −2 and ended up even more negative at −8. |

7. Add or subtract. Think of the number line jumps.

| a. 3 − 4 =

2 − 5 =

5 − 9 = | b. ⁻2 − 1 =

⁻6 − 4 =

⁻7 − 2 = | c. ⁻4 + 4 =

⁻7 + 3 =

⁻12 + 5 = | d. ⁻5 + 6 =

⁻8 + 4 =

⁻6 + 7 = |

8. Find the number that is missing from the equations. Think of moving on the number line.

| a. 1 − _____ = ⁻4

b. 3 − _____ = ⁻3 | c. ⁻7 + _____ = ⁻6

d. ⁻9 + _____ = ⁻1 | e. 2 − _____ = ⁻5

f. 0 − _____ = ⁻8 | g. ⁻3 + _____ = 0

h. ⁻9 + _____ = 9 |

9. The expression 1 − 2 − 3 − 4 can also be thought of as a person making jumps on the number line. Where does the person end up?

10. James had $5. He bought coffee for $2 and a sandwich for $6. He paid what he could and the cashier put the rest of what he owed on his charge account with the cafe.

   a. Write a math sentence to show the transaction.

   b. How much in debt is James now?

What about adding or subtracting a negative number?

Here is a way to think about 3 − (−2). Imagine you are standing at 3 to start with. Because of the subtraction sign, you turn to the left and get ready to take your steps. However, because of the additional minus sign in front of the 2, you have to take those steps BACKWARD—to the right! So, because you ended up taking those 2 steps to the *right*, in effect, you have just performed 3 + 2.

Another example: Here is a way to think about −4 + (−5). Imagine you are standing at −4 to start with. Because of the addition sign, you turn to the right and get ready to move. But because of the additional minus sign in front of the 5, you have to take those 5 steps BACKWARD. So you take those 5 steps to the *left* instead. In essence, you have performed −4 − 5. (In the next lesson we will examine *other* ways to think about these situations.)

Adding Integers: Counters

Addition of integers can be modeled using **counters**. We will use green counters with a "+" sign for positives and red counters with a "−" sign for negatives.

  Here we have the sum 2 + 3. There is a group of 2 positives and another of 3 positives.	 This picture shows the sum (−2) + (−3). We *add* negatives and negatives. In total, there are five negatives, so the sum is −5.	 1 + (−1) = 0 One positive counter and one negative counter *cancel* each other. In other words, their sum is zero!
  2 + (−2) = 0 Two negatives and two positives also cancel each other. Their sum is zero.	 3 + (−1) = 2 Here, one "positive-negative" pair is cancelled (you can cross it out!). We are left with 2 positives.	 (−4) + 3 = −1 Now the negatives outweigh the positives. Pair up three negatives with three positives. Those cancel out. There is still one negative left.

1. Refer to the pictures and add. Remember each "positive-negative" pair is cancelled.

  a. 2 + (−5) = _____	 **b.** (−3) + 5 = _____	 **c.** (−6) + (−3) = _____
  d. 3 + (−5) = _____	 **e.** 2 + (−4) = _____	 **f.** (−8) + 5 = _____

2. Write addition sentences (equations) to match the pictures.

  a.	 **b.**	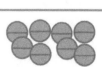 **c.**
  d.	 **e.**	 **f.**

> **A note on notation**
>
> We can write an elevated minus sign to indicate a negative number: ⁻4.
> Or we can write it with a minus sign and brackets: (−4).
> We can even write it without the brackets if the meaning is clear: −4.
>
> So ⁻4 + ⁻4 = ⁻8 is the same as (−4) + (−4) = (−8), which is the same as −4 + (−4) = −8.

> You *should* write the brackets if you have + and −, or two − signs, next to each other.
>
> So, do *not* write "8 + − 4"; write "8 + (−4)." And do not write "3 − −3"; write "3 − (−3)."

3. Think of the counters. Add.

a. 7 + (−8) = (−7) + 8 =	b. (−7) + (−8) = 7 + 8 =	c. 5 + (−7) = 7 + (−5) =	d. 50 + (−20) = 10 + (−40) =
e. ⁻2 + ⁻4 = ⁻6 + 6 =	f. 10 + ⁻1 = ⁻10 + ⁻1 =	g. ⁻8 + 2 = ⁻8 + ⁻2 =	h. ⁻9 + ⁻1 = 9 + ⁻1 =

4. Rewrite these sentences using symbols, and solve the resulting sums.

   a. The sum of seven positives and five negatives.

   b. Add −3 and −11.

   c. Positive 100 and negative 15 added together.

5. Write a sum for each situation and solve it.

   a. Your checking account is overdrawn by $50. (This means your account is negative).
   Then you earn $60. What is the balance in your account now?

   b. Hannah owed her mother $20. Then, she borrowed $15 more from her mother.
   What is Hannah's "balance" now?

6. Consider the four expressions 2 + 6, (−2) + (−6), (−2) + 6 and 2 + (−6). Write these expressions in order from the one with **least** value to the one with **greatest** value.

7. Find the number that is missing from the equations.

a. −3 + ____ = −7	b. −3 + ____ = 3	c. 3 + ____ = (−7)
d. ____ + (−15) = −22	e. 2 + ____ = −5	f. ____ + (−5) = 0

> **Comparing number line jumps and counters**
>
> We can think of $-5 + (-3)$ as five negatives and three negatives, totaling 8 negatives or -8. We also know that $-5 - 3$ is like starting at -5 and jumping three steps towards the left on the number line, ending at -8.
>
> Since both have the same answer, the two expressions $-5 + (-3)$ and $-5 - 3$ are equal:
>
> $$-5 + (-3) = -5 - 3$$
>
> It is as if the "+ −" in the middle is changed into a single − sign. This, indeed, is a *shortcut*!
>
> Similarly, $2 + (-7)$ is the same as $2 - 7$. Either (1) think of having 2 positive and 7 negative counters, totaling 5 negatives, (2) or think of being at 2 and taking 7 steps to the left, ending at -5.
>
> **With integer problems, you can think of number line jumps or of counters, whichever is easier.**

8. Compare how $-7 + 4$ is modeled on the number line and with counters.

   a. On the number line, $-7 + 4$ is like starting at _____, and moving _____ steps to the _____,
   ending at _____.

   b. With counters, $-7 + 4$ is like _____ negatives and _____ positives added together. We can
   form _____ negative-positive pairs that cancel, and what is left is _____ negatives.

9. Add.

a. $4 + (-10) =$	**b.** $-8 + (-8) =$	**c.** $-5 + (-7) =$	**d.** $11 + (-2) =$
$-6 + 8 =$	$7 + (-8) =$	$12 + (-5) =$	$-10 + 20 =$

10. **a.** Find the value of the expression $x + (-4)$ for four different values of x.
    You can choose the values.

    b. For which value of x does the expression $x + (-4)$ have the value 0?

11. Solve the problems, and observe the patterns.

a. $3 - 2 =$	**b.** $^-7 - 0 =$	**c.** $^-5 + 0 =$	**d.** $^-6 + 6 =$
$3 - 3 =$	$^-7 - 1 =$	$^-5 + 1 =$	$^-6 + 7 =$
$3 - 4 =$	$^-7 - 2 =$	$^-5 + 2 =$	$^-6 + 8 =$
$3 - 5 =$	$^-7 - 3 =$	$^-5 + 3 =$	$^-6 + 9 =$
$3 - 6 =$	$^-7 - 4 =$	$^-5 + 4 =$	$^-6 + 10 =$

Subtracting a Negative Integer

We have already looked at such subtractions as 3 − 5 or −2 − 8, which you can think of as number line jumps. But what about **subtracting a negative integer?** What is 5 − (−4)? Or (−5) − (−3)?

Let's look at this kind of expression with a "double negative" in several different ways.

1. Subtraction as "taking away":

We can model subtracting a negative number using counters. (−5) − (−3) means we start with 5 negative counters, and then we *take away* 3 negative counters. That leaves 2 negatives, or −2.

5 − (−4) cannot easily be modeled that way, because it is hard to take away 4 negative counters when we do not have any negative counters to start with. But you *could* do it this way:

Start out with 5 positives. Then *add* four positive-negative pairs, which is just adding zero! Now you can take away four negatives. You are left with nine positives.

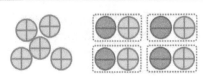

Start out with 5.   Add four positive-negative pairs, which amount to zero.

Lastly, cross out four negatives. You are left with nine positives.

2. Subtracting a negative number as a number line jump:

5 − (−4) is like standing at 5 on the number line, and getting ready to subtract, or go to the left. But, since there is a minus sign in front of the 4, it "turns you around" to face the positive direction (to the right), and you take 4 steps to the right instead. So, 5 − (−4) = 5 + 4 = 9.

(−5) − (−3) is like standing at −5, ready to go to the left, but the minus sign in front of 3 turns you "about face," and you take 3 steps to the right instead. You end up at −2.

3. Subtraction as a difference/distance:

To find the difference or distance between 76 and 329, subtract 329 − 76 = 253 (the smaller-valued number from the bigger-valued one). If you subtract the numbers the other way, 76 − 329, the answer is −253.

By the same analogy, we can think of 5 − (−4) as meaning the difference or distance between 5 and −4. From the number line we can see the distance is **9**.

(−5) − (−3) *could* be the distance between −5 and −3, except it has the larger number, −3, subtracted from the smaller number, −5.

If we turn them around, (−3) − (−5) would give us the distance between those two numbers, which is 2. Then, (−5) − (−3) would be the opposite of that, or −2.

Two negatives make a positive!

You have probably already noticed that, any way you look at it, we can, in effect, replace those two minuses in the middle with a + sign. In other words, 5 − (−4) has the same answer as 5 + 4. And (−5) − (−3) has the same answer as −5 + 3. It may look a bit strange, but it works out really well.

$5 - (-4)$
$5 + 4 = 9$

$(-5) - (-3)$
$(-5) + 3 = -2$

1. Write a subtraction sentence to match the pictures.

   a.

   b.

2. Write an addition or subtraction sentence to match the number line movements.

   a. You are at −2. You jump 6 steps to the left.

   b. You are at −2. You get ready to jump 6 steps to the left,
   but turn around at the last minute and jump 6 steps to the right instead.

3. Find the distance between the two numbers. Then, write a matching subtraction sentence. To get a positive distance, remember to *subtract the smaller number from the bigger number.*

a. The distance between 3 and −7 is _____.	**b.** The distance between −3 and −9 is _____.
Subtraction: _____ − _____ = _____	Subtraction: _____ − _____ = _____
c. The distance between −2 and 10 is _____.	**d.** The distance between −11 and −20 is _____.
Subtraction: _____ − _____ = _____	Subtraction: _____ − _____ = _____

4. Solve. Remember the shortcut: you can change each double minus "− −" into a plus sign.

a. −8 − (−4) =	**b.** −1 − (−5) =	**c.** 12 − (−15) =
8 − (−4) =	1 − (−5) =	−12 + 15 =
−8 + (−4) =	−1 − 5 =	−12 − 15 =
8 + (−4) =	1 − 5 =	12 + (−15) =

5. Connect with a line the expressions that are equal (have the same value).

a.		b.	
10 − (−3) 10 − 3		−9 + 2 −9 + (−2)	
10 + (−3) 10 + 3		−9 − 2 −9 − (−2)	

6. Write an integer addition or subtraction to describe the situations.

   a. A roller coaster begins at 90 ft above ground level.
   Then it descends 105 feet.

   b. Matt has $25. He wants to buy a bicycle from his friend that costs $40.
   How much will he owe his friend?

Solve −1 + (−2) − (−3) − 4.

Puzzle Corner

Add and Subtract Roundup

1. Addition of integers.

- Think of positive and negative counters.
- When adding a negative and positive integer, you can also think of number line jumps.

Example 1. $-4 + 11$

a. Think of a number line: you are at -4 and jump 11 steps to the right. You end up at 7.

b. Think of 4 negative and 11 positive counters. Four negative-positive pairs cancel each other, and 7 positives are left.

Example 2. $-4 + (-11)$

Think of 4 negative and 11 negative counters. The negatives add up, giving us -15.

Example 3. $4 + (-11)$

Think of 4 positive and 11 negative counters. Four negative-positive pairs cancel each other, and 7 negatives are left.

1. Addition skills check-up! Add.

a. $-4 + 6 =$ _____	b. $-3 + 7 =$ _____	c. $8 + (-9) =$ _____
$4 + (-6) =$ _____	$-3 + (-7) =$ _____	$(-8) + (-9) =$ _____

2. Subtraction of integers.

- When subtracting a positive integer, think of number line jumps.
- When subtracting a negative integer, use the shortcut where the double negative becomes a positive, and the subtraction turns into an addition.

Example 4. $-4 - 11$

Think of a number line jump: you are at -4 and jump 11 steps to the left. You end up at -15.

Example 5. $4 - 11$

Think of a number line jump: you are at 4 and jump 11 steps to the left. You end up at -7.

Example 6. $-4 - (-11)$

Turn the double negative into a positive: $-4 - (-11) = -4 + 11$. Now think of the counters or a number line jump. The answer is 7.

Example 7. $4 - (-11)$

Turn the double negative into a positive: $4 - (-11) = 4 + 11 = 15$.

2. Subtraction skills check-up! Subtract.

a. $(-4) - 6 =$ _____	b. $-3 - 7 =$ _____	c. $8 - (-9) =$ _____
$4 - (-6) =$ _____	$-3 - (-7) =$ _____	$(-8) - (-9) =$ _____

3. Find the missing integer.

a. $-4 +$ _____ $= -10$	b. $6 +$ _____ $= 0$	c. $5 -$ _____ $= -2$
$4 +$ _____ $= -2$	$-6 -$ _____ $= -4$	$4 +$ _____ $= 1$

4. Write an addition or subtraction sentence to match the number line jumps.

   a. You are at ⁻6. You jump 5 to the right. You end up at _____.

   b. You are at ⁻2. You jump 7 to the right. You end up at _____.

   c. You are at 4. You jump 3 to the left. You end up at _____.

   d. You are at 0. You jump 12 to the left. You end up at _____.

   e. You are at 7. You jump 22 to the left. You end up at _____.

   f. You are at ⁻7. You jump 22 to the right. You end up at _____.

5. Write an addition or subtraction sentence with integers to match these situations.

Situation	Addition or Subtraction Sentence
A miner was in an underground elevator, 65 m below the surface. The elevator descended 35 m. Now it is at an elevation of _____ m.	
Henry paid $150 for car repairs using a credit card (in other words, he made a debt). He paid $60 of the debt. Then he used the credit card again for another $120 of expenses. Now Henry owes $_____ on his credit card.	
Amy owed $180. She paid $40 of her debt. Her money situation is now _____.	
The temperature was 2°C. During the night it dropped 5 degrees. At dawn, it rose 2 degrees. Now the temperature is _____°C.	

6. Here is a funny riddle. Solve the math problems to uncover the answer.

   L ____ + (−8) = −12   **R** −24 − (−25) = _____   **M** (−144) + 150 = ____

   O 3 + (−11) = ____   **A** ____ + 4 = 2   **H** 77 − 90 = ____

   C −2 − ____ = −5   **I** ____ + (−7) = 0   **I** −4 + (−15) _____

   S −2 − (−4) = _____   **O** −3 + 5 + (−5) = _____   **C** −7 + ____ = 2

   S −2 − ____ = 5   **A** 7 − ____ = −1

   What game do cows play at parties?

   | 6 | −8 | −3 | 2 | −19 | 9 | 8 | −4 | | 3 | −13 | −2 | 7 | 1 | −7 |

   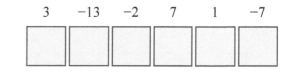

Graphing

Remember? When an equation has two variables, there are many values of x and y that make that equation true.

Example. Note the equation $y = 2 - x$. If $x = 0$, then we can <u>calculate</u> the value of y using the equation: $y = 2 - 0 = 2$.

So, when $x = 0$ and $y = 2$, that equation is true. We can plot the number pair (0, 2) on the coordinate grid.

Some of the other (x, y) values that make the equation true are listed below, and they are plotted on the right.

x	−4	−3	−2	−1	0	1	2	3	4
y	6	5	4	3	2	1	0	−1	−2

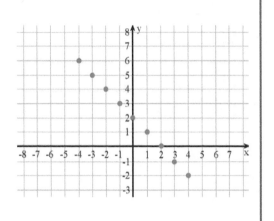

1. Plot the points from the equations. Graph both (b) and (c) in the same grid.

   a. $y = x + 4$

x	−9	−8	−7	−6	−5	−4	−3	−2
y								

x	−1	0	1	2	3	4	5	6
y								

   b. $y = 6 - x$

x	−3	−2	−1	0	1	2	3
y							

x	4	5	6	7	8	9
y						

   c. $y = x - 2$

x	−5	−4	−3	−2	−1	0	1	2
y								

x	3	4	5	6	7	8	9
y							

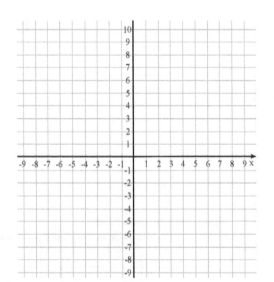

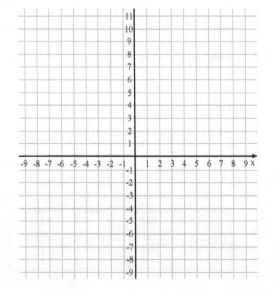

2. Continue the patterns according to the given rules, and then plot the points using the number pairs.

   a.

add 1	x	−5	−4	−3	−2	−1
add 2	y	−8	−6	−4		

add 1	x	0	1	2	3	4
add 2	y					

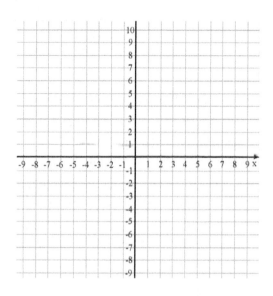

   b.

add 2	x	−8	−6	−4		
subtract 3	y	9	6			

add 2	x					
subtract 3	y					

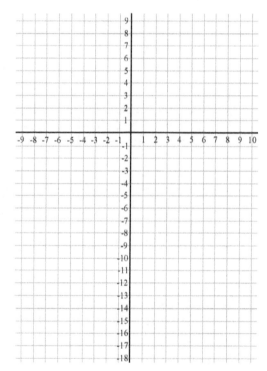

   c. Make up your own rules.

	x					
	y					

	x					
	y					

   Make another if you would like!

	x					
	y					

	x					
	y					

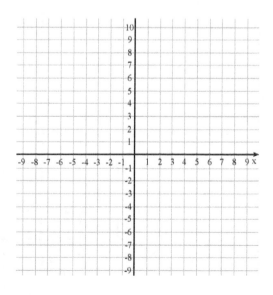

3. On 1 March, Natalie decided she would start to pay off her debt of $200. She decided she would pay $15 each week. Let t denote the number of weeks since 1 March. We can write in a table Natalie's progress with the payments. The variable a denotes her balance.

   a. Fill in the table.

t	0	1	2	3	4	5	6	7	8	9	10	11	12	13
a	−200	−185	−170	−155	−140	−125	−110	−95	−80	−65	−50	−35	−20	−5

   b. Plot these points on the coordinate grid below.

   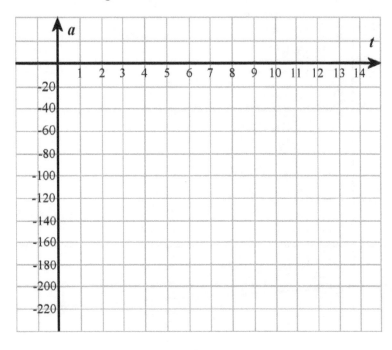

   c. When will Natalie finish paying off her debt?

   After 14 weeks (during the 14th week, since $200 \div 15 \approx 13.33$).

   d. How can you see that from the graph?

   The line (points) crosses the t-axis (where $a = 0$) between $t = 13$ and $t = 14$.

4. **a.** Write the points from the graph in the table.

x	−2	−1	1	2	3	4	5
y	−7	−5	−1	1	3	5	7

   b. Find the pattern the x- and y-coordinates follow in (a), and use that same pattern to fill in the table below.

x	−8	−7	−6	−5	−4	−3	−2
y	−19	−17	−15	−13	−11	−9	−7

   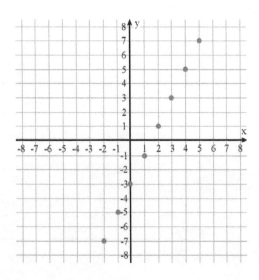

5. Sofia has $500 in her bank account on 1 January. Out of her bank account comes an automatic payment of $35, once a month, for Internet service. Let's say Sofia does not use this bank account for anything else. Let t be the number of months since January (January would be month zero).

   a. Fill in the table for Sofia's balance.

months	t	0	1	2	3	4	5	6	7	8	9	10	11	12
balance	a	$500												

   b. Plot the points.

   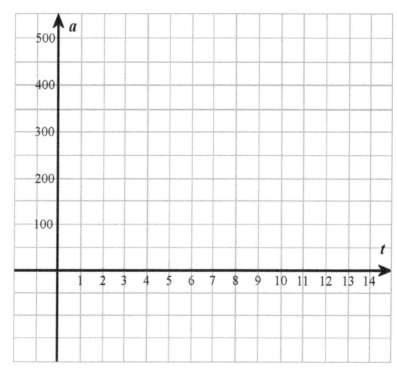

   c. If Sofia does not add more funds to her account, when will the account be negative?

   d. Let's say Sofia *forgets* to add more funds to her bank account. What is her balance at 14 months?

Puzzle Corner

x	0	1	2	3	4
y	6	4	2	0	−2

Look at the pattern that relates x and y.

Continuing on with the same pattern, what would y be when x is 100?

Chapter 8 Mixed Review

1. Write an equation for each situation EVEN IF you could easily solve the problem without an equation! Then solve the equation.   (Writing Equations/Ch.2)

   a. Katie is 54 years old. Shelly is 12 years younger than Katie. How old is Shelly?

   b. Bob bought some tulips for his wife. One tulip cost $2.15. The total cost was $45.15. How many tulips did Bob buy?

2. Find a number between 500 and 520 whose prime factorization has only 2s.
   (The Sieve of Eratosthenes and Prime Factorization/Ch.6)

3. Write either a multiplication or a division, and solve.   (Dividing Fractions: Reciprocal Numbers/Ch.7)

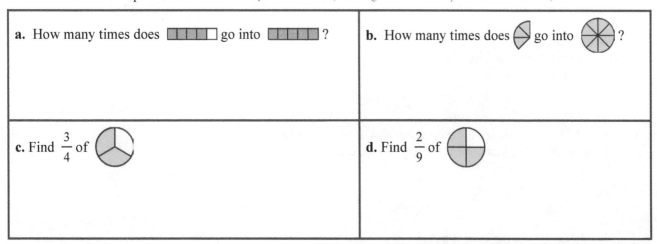

4. A package of cheap dominoes weighs 3 oz. A package of quality dominoes weighs 1 lb.
   (Convert Customary Measuring Units/Ch.3)

   a. How much does a box containing 50 packages of the cheap dominoes weigh? Give your answer in kilograms.

   b. Another box contains 24 packages of the quality dominoes. Find how much more the box with quality dominoes weighs than the box with cheap dominoes.

5. Divide. (Divide Fractions/Ch.7)

a. $2\frac{7}{8} \div \frac{2}{5}$	**b.** $4 \div 1\frac{5}{6}$
c. $5 \div \frac{2}{7}$	**d.** $10\frac{1}{10} \div \frac{3}{4}$

6. Annabelle can type 70 words in two minutes. How many words can she type in 15 minutes?
   (Unit Rates *and* Using Equivalent Rates/Ch.4)

7. Mother's and Father's ages are in the ratio of 7:8. Father is six years older than Mother. How old is Mother?
   (Ratio Problems and Bar Models 2/Ch.4)

8. A rectangle's aspect ratio is 5:2, and its perimeter is 84 cm. Find its area.
   (Aspect Ratio/Ch.4)

9. Keith paid $414 of his salary in taxes. After that, he had $1,459 left. What percentage of his income did Keith pay in taxes?
   (What Percentage...?/Ch.5)

10. Express these rates in the lowest terms. (Unit Rates/Ch.4)

a. 720 km : 4 hr	**b.** 6 kg for $4.20	**c.** 120 miles on 5 gallons

11. Simplify before you multiply. (Review: Multiplying Fractions 1/Ch.7)

a. $\frac{5}{36} \cdot \frac{24}{45}$	**b.** $\frac{16}{30} \cdot \frac{25}{24}$	**c.** $\frac{14}{25} \cdot \frac{35}{42}$

Integers Review

1. Compare. Write < or > in the box.

   a. −1 ☐ −7   b. 2 ☐ −2   c. −6 ☐ 0   d. 8 ☐ −3   e. −8 ☐ −3

2. Order the numbers from the least to the greatest.

   a. −6   2   −2   0

   b. −14   −8   −11   −7

3. Express the situations using integers. Then compare them writing > or < in the box.

a. Lillian owes $12 and Haley owes $18.	____ ☐ ____
b. At 2 PM, the temperature was 5°C below zero. Now it is 2°C.	____ ☐ ____
c. Joe rose in an elevator to the height of 16 m, whereas Gabriel went down 6 m below the ground.	____ ☐ ____

4. Simplify. In (e), write using a number.

   a. $|-11|$ b. $|2|$ c. $|0|$ d. $-(-19)$ e. the opposite of 7

5. Draw a number line jump for each addition or subtraction sentence.

   a. $-9 + 6 =$ _____   b. $-2 + 5 =$ _____

   ````
   +--+--+--+--+--+--+--+--+--+--+--+--+--+--+
   -10 -9 -8 -7 -6 -5 -4 -3 -2 -1  0  1  2  3
   ````

   c. $-3 - 5 =$ _____   d. $2 - 8 =$ _____

   ````
   +--+--+--+--+--+--+--+--+--+--+--+--+--+--+
   -10 -9 -8 -7 -6 -5 -4 -3 -2 -1  0  1  2  3
   ````

6. Write an addition or subtraction sentence.

   a. You are at ⁻10. You jump 6 to the right. You end up at _____.

   b. You are at ⁻5. You jump 8 to the right. You end up at _____.

   c. You are at 3. You jump 7 to the left. You end up at _____.

   d. You are at ⁻11. You jump 3 to the left. You end up at _____.

7. Add or subtract.

a.	b.	c.	d.
2 + (−8) = _____	−2 + (−9) = _____	1 + (−7) = _____	5 − (−2) = _____
(−2) + 8 = _____	2 − 8 = _____	−4 − 5 = _____	−3 − (−4) = _____

8. Write an addition or a subtraction sentence to match the situations.

   a. May has $35. She wants to purchase a guitar for $85.
      That would make her money situation be _____.

   b. A fish was swimming at the depth of 6 ft. Then he sank 2 ft.
      Then he sank 4 ft more. Now he is at the depth of _____ ft.

   c. Elijah owed his father $20. Then he borrowed another $10.
      Now his balance is _____.

   d. The temperature was −13°C and then it rose 5°.
      Now the temperature is _____ °C.

9. Use mathematical symbols to express these ideas

   a. the distance of −17 from zero

   b. the opposite of −11

10. Which expression below matches with the situation? Jacob owes more than fifty dollars.

   a. balance > −$50   b. balance = −$50   c. balance < $50   d. balance < −$50

11. Plot the points from the function $y = 4 - x$
    for the values of x listed in the table.

x	−5	−4	−3	−2	−1	0	1	2
y								

x	3	4	5	6	7	8	9
y							

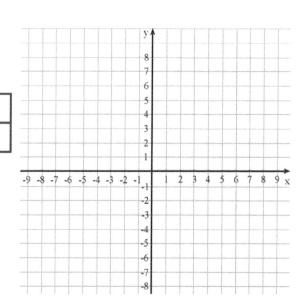

12. Find the missing integers.

a. −2 + _____ = −8	b. 4 + _____ = 0	c. 5 − _____ = −2
3 + _____ = −2	−6 − _____ = −12	3 + _____ = 1

13. Andrew drew a polygon, and then he reflected it in the x-axis. The vertices of the reflected polygon are: (−9, 6), (−6, 6), (−9, 3) and (−3, 0). What were the coordinates of the original vertices?

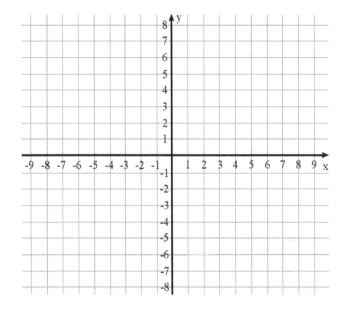

14. Find the distance between the points.

   a. (−3, −12) and (−3, 15)

   b. (−15, 34) and (−21, 34)

15. a. The points (−7, −3), (−1, −7), (−1, −1) and (−4, −6) are vertices of a quadrilateral. Draw the quadrilateral.

   b. Reflect the quadrilateral in the y-axis. (Draw the new quadrilateral). Write the coordinates of the moved vertices.

   (−7, −3) → (_____ , _____)

   (−1, −7) → (_____ , _____)

   (−1, −1) → (_____ , _____)

   (−4, −6) → (_____ , _____)

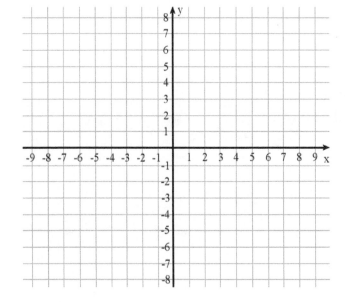

   c. Now move the already reflected quadrilateral 7 units up. (Draw the new quadrilateral). Write the coordinates of the new vertices.

   (_____ , _____)   (_____ , _____)   (_____ , _____)   (_____ , _____)

Chapter 9: Geometry
Introduction

The main topics in this chapter include:

- area of triangles
- area of polygons
- nets and surface area of prisms and pyramids
- volume of rectangular prisms with sides of fractional length

However, the chapter starts out with some review of topics from earlier grades, as we review the different types of quadrilaterals and triangles and students do some basic drawing exercises. In these drawing problems, students will need a ruler to measure lengths and a protractor to measure angles.

One focus of the chapter is the area of polygons. To reach this goal, we follow a step-by-step development. First, we study how to find the area of a right triangle, which is very easy, as a right triangle is always half of a rectangle. Next, we build on the idea that the area of a parallelogram is the same as the area of the related rectangle, and from that we develop the usual formula for the area of a parallelogram as the product of its base times its height. This formula then gives us a way to generalize finding the area of any triangle as *half* of the area of the corresponding parallelogram.

Finally, the area of a polygon can be determined by dividing it into triangles and rectangles, finding the areas of those and summing them. Students also practice their new skills in the context of a coordinate grid. They draw polygons in the coordinate plane and find the lengths of their sides, perimeters and areas.

Nets and surface area is another major topic. Students draw nets and determine the surface area of prisms and pyramids using nets. They also learn how to convert between different area units, not using conversion factors or formulas, but using logical reasoning where they learn to determine those conversion factors themselves.

Lastly, we study the volume of rectangular prisms, this time with edges of fractional length. (Students have already studied this topic in fifth grade with edges that are a whole number long.) The basic idea is to prove that the volume of a rectangular prism *can* be calculated by multiplying its edge lengths even when the edges have fractional lengths. To that end, students need to think how many little cubes with edges ½ or ⅓ unit go into a larger prism. Once we have established the formula for volume, students solve some problems concerning the volume of rectangular prisms.

There are quite a few videos available to match the lessons in this chapter at
https://www.mathmammoth.com/videos/ (choose 6th grade).

Also, don't forget to use the resources for challenging problems:
https://l.mathmammoth.com/challengingproblems

I recommend that you at least use the first resource listed, Math Stars Newsletters.

The Lessons in Chapter 9

	page	span
Quadrilaterals Review	105	*3 pages*
Triangles Review	108	*2 pages*
Area of Right Triangles	110	*2 pages*
Area of Parallelograms	112	*3 pages*

Area of Triangles	115	*2 pages*
Polygons in the Coordinate Grid	117	*3 pages*
Area of Polygons	120	*2 pages*
Area of Shapes Not Drawn on Grid	122	*2 pages*
Area and Perimeter Problems	124	*2 pages*
Nets and Surface Area 1	126	*3 pages*
Nets and Surface Area 2	129	*2 pages*
Problems to Solve – Surface Area	131	*2 pages*
Converting Between Area Units	133	*2 pages*
Volume of a Rectangular Prism with Sides of Fractional Length	135	*3 pages*
Volume Problems	138	*2 pages*
Chapter 9 Mixed Review	140	*3 pages*
Geometry Review	143	*3 pages*

Helpful Resources on the Internet

We have compiled a list of Internet resources that match the topics in this chapter. This list of links includes web pages that offer:

- **online practice** for concepts;
- online **games**, or occasionally, printable games;
- **animations** and interactive **illustrations** of math concepts;
- **articles** that teach a math concept.

We heartily recommend you take a look at the list. Many of our customers love using these resources to supplement the bookwork. You can use the resources as you see fit for extra practice, to illustrate a concept better, and even just for some fun. Enjoy!

https://l.mathmammoth.com/gr6ch9

Quadrilaterals Review

Review the definitions of various quadrilaterals below.

- A **rectangle** has four right angles.
- A **square** is a rectangle with four congruent sides.
- A **trapezoid** has *at least* one pair of parallel sides. It may have two!
- A **parallelogram** has two pairs of parallel sides.
- A **rhombus** is a parallelogram that has four congruent sides (a diamond).
- A **kite** has two pairs of congruent sides that touch each other. *The single tick marks show the one pair of congruent sides, and the double tick marks show the other pair.*
- In a **scalene** quadrilateral, all sides are of different lengths (no two sides are congruent).

The chart shows the seven different types of quadrilaterals as a "family," descending from the generic quadrilateral at the top.

If a quadrilateral is listed under another, it means the two have like a "parent-child" relationship: the quadrilateral listed lower (the child) has its parent's characteristics.

Number in the chart the following types of quadrilaterals:

1. rhombus
2. kite
3. rectangle
4. square
5. scalene quadrilateral
6. trapezoid
7. parallelogram

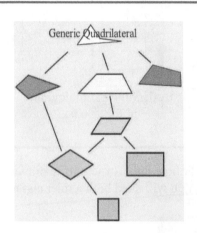

1. Match each description to a name of a quadrilateral.

- All of the sides measure 6 cm; angles measure 50°, 130°, 50° and 130°.
- Two of the sides measure 12 cm and two measure 8 cm; angles measure 60°, 120°, 60° and 120°; opposite sides are parallel.
- None of the sides are congruent.
- The sides, listed in order, measure 70 cm, 70 cm, 120 cm and 120 cm.
- The sides measure 20 cm, 35 cm, 20 cm, and 55 cm. The two longest sides are parallel.
- All of the sides measure 16 cm; the angles all measure 90°.

square
scalene quadrilateral
trapezoid
rhombus
kite
parallelogram

2. Find the correct type of quadrilateral for each definition.

   a. A quadrilateral with four congruent sides (but we know nothing about its angles).

   b. A quadrilateral where the opposite sides are parallel (that is all we know about it!).

   c. A quadrilateral with one pair of parallel sides.

   d. A quadrilateral with two pairs of congruent sides, where the two congruent sides are touching (adjacent), and also the other two congruent sides are touching.

   e. A quadrilateral with four congruent sides and four right angles.

3. Think of the "parent-child" relationships as shown in the chart, and answer the questions:

   a. Is a rhombus also a kite?

   b. Is a rectangle also a parallelogram?

   c. Is a trapezoid also a rectangle?

4. Reflect the parallelograms in the coordinate grid.

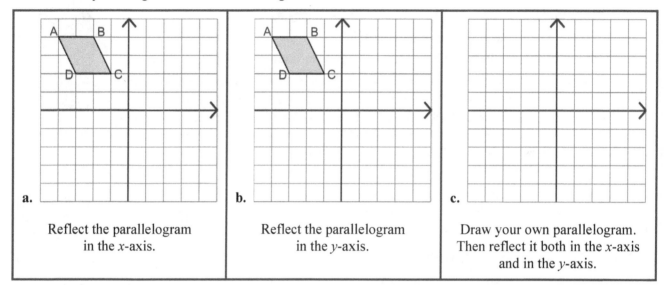

a. Reflect the parallelogram in the *x*-axis.

b. Reflect the parallelogram in the *y*-axis.

c. Draw your own parallelogram. Then reflect it both in the *x*-axis and in the *y*-axis.

5. Follow the instructions to draw a rhombus with 6-cm sides and 30° and 150° angles. You will need both a ruler and a protractor.

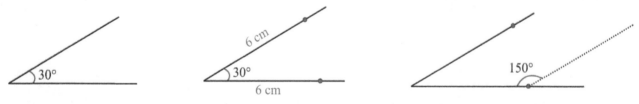

(i) Draw a 30° angle.

(ii) Continue the two sides of your angle so you can mark off the two 6-cm sides. The points you mark are two other vertices of the rhombus.

(iii) Now draw a 150° angle a vertex you marked at step (2).

(iv) Figure out the last step, and finish drawing your rhombus.

6. Follow the instructions to draw a parallelogram with a 10-cm and 6-cm sides and a 60° and 120° angles.

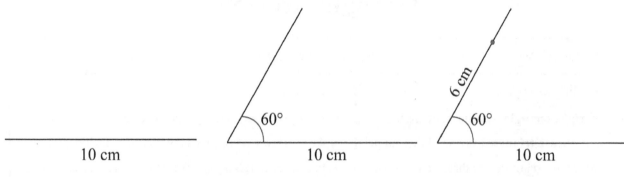

(i) Draw a 10-cm side.   (ii) Draw the 60° angle.   (iii) Now mark off the 6-cm side.

(iv) Figure out the next steps, and finish drawing your parallelogram.

7. Draw a trapezoid where the parallel sides measure 4 cm and 7 cm. Use the proper tools (a ruler in this case).

Is there only one such trapezoid, or could you draw several different ones that would match the description?

Hint: start by drawing the 7-cm side.

Triangles Review

Remember, we can classify a triangle both according to its <u>angles</u> and according to its <u>sides</u>.	
According to angles:	According to sides:
• A **right triangle** has one right angle.	• An **equilateral triangle** has three congruent sides.
• An **obtuse triangle** has one obtuse angle.	• An **isosceles triangle** has two congruent sides.
• An **acute triangle** has three acute angles.	• In a **scalene triangle**, none of the sides are congruent.

1. Match the classifications and the triangles.

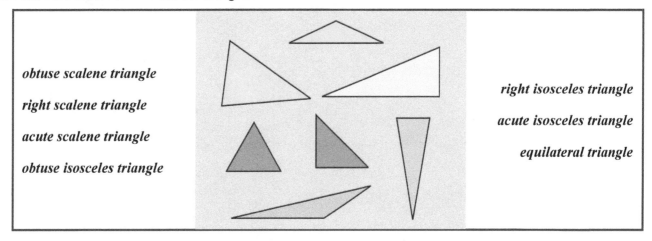

obtuse scalene triangle

right scalene triangle

acute scalene triangle

obtuse isosceles triangle

right isosceles triangle

acute isosceles triangle

equilateral triangle

2. Which are impossible?

    isosceles scalene triangle   isosceles acute triangle   isosceles obtuse triangle

                  scalene right triangle   equilateral obtuse triangle

3. **a.** Draw an isosceles triangle with a 78° top angle and two 7.6-cm sides. Start out by drawing the 78° angle, then measure the two 7.6-cm sides.

   b. Measure the base angles of your triangle. They should be congruent; however, it is very hard to draw so accurately that you would get the same angle measurements for the base angles, so don't worry if they differ a little bit.

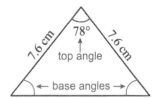

These two pictures illustrate how it is ENOUGH to know the measurements of two angles and the length of the side *between* them in order to draw a triangle.

In other words, you do not have to know *all* the angles and the sides in order to draw a triangle —just two angles and a side is sufficient.

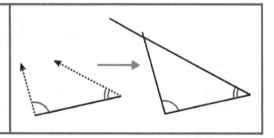

4. **a.** Draw a triangle that has a 25° angle, a 115° angle and a 8-cm side between those angles.

   b. Classify it according to its sides and its angles.

5. **a.** Draw a triangle that has a 67° angle, a 75° angle and a 5-cm side between those angles.

   b. Classify it according to its sides and its angles.

6. Draw a copy of this triangle. Your triangle should match this triangle exactly if they were placed on top of each other. Hint: first measure some angle(s) and some side(s).

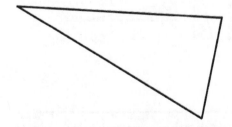

Area of Right Triangles

This rectangle is divided into two right triangles that are **congruent.** This means that if you could flip one of them and move it on top of the other, they would match exactly.

The rectangle has an area of 2 · 4 = 8 square units.
Can you figure out what the area of just *one* of the triangles is?

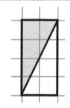

Here the area of the whole rectangle is 3 · 5 = 15 square units.
How could you figure out the area of just one of the triangles?

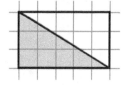

Here the sides of the triangle are 6 and 3 units. The other two sides of the rectangle are drawn with dotted lines. The area of the *rectangle* is 18 square units. The area of just the triangle is half of that, or 9 square units.

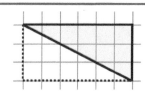

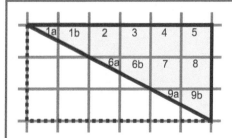

Let's look closer at the last triangle above. To confirm that its area is 9 square units, we can *count* the little squares in the triangle.

Notice that some of the parts do not cover a complete square, but by combining those we can make whole squares and then count them.

1. Find the area of these right triangles. To help you, trace the "helping rectangle" for the triangles.

a. _____ square units

b. _____ square units

c. _____ square units

d. _____ square units

e. _____ square units

f. _____ square units

g. _____ square units

h. _____ square units

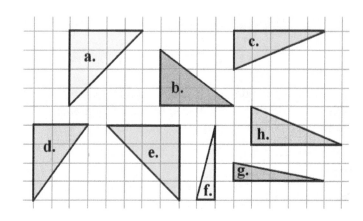

To find the area of a right triangle, **multiply the lengths of the two sides** that are *perpendicular* to each other (in other words, the two that form the right angle). Then **take half of that**.

This works because the area of the right triangle is exactly _____ of the area of the rectangle.

110

2. Draw a right triangle whose two perpendicular sides are given below, and then find its area.

   a. 4 cm and 7 cm

   b. 6.5 cm and 3 cm

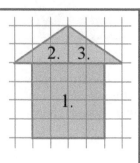

We can find the area of this house-shape in three parts.

1. The square has an area of 4 · 4 = 16 square units.

2. Triangle 2 has perpendicular sides of 3 and 2 units, so its area is (1/2) · 2 · 3 = 3 square units.

3. Triangle 3 is the same shape and size as triangle 2, so its area is also 3 square units.

Lastly, add the areas: 16 + 3 + 3 = 22 square units in total.

3. Find the areas of these compound shapes.

a. 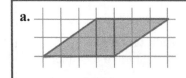 b. 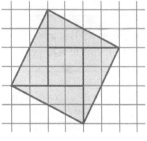 c.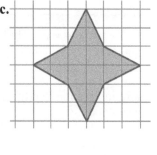

4. Draw a right triangle whose *area* is 15 square centimeters.
   Can you only draw one right triangle with that area, or several different kinds?
   Explain.

5. In the grid, draw 3 different right triangles that each have an area of 6 square units.

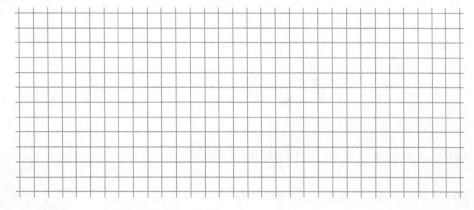

111

Area of Parallelograms

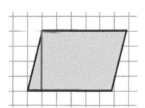 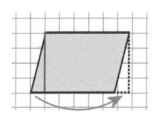 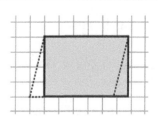

We draw a line from one vertex of the parallelogram in order to form a *right triangle*. Then we move the triangle to the other side, as shown. Look! We get a *rectangle*!

The rectangle's area is 6 · 4 = 24 square units, and that is *also the area of the original parallelogram*.

It works here, as well. The area of the rectangle and of the parallelogram are the same: both have the area of 4 · 4 = 16 square units.

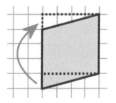 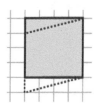

The area of a parallelogram is the same as the area of the corresponding rectangle.
You construct the rectangle by moving a right triangle from one side of the parallelogram to the other.

1. Imagine moving the marked triangle to the other side as shown. What is the area of the original parallelogram?

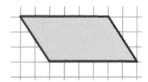   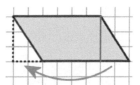

2. Draw a line in each parallelogram to form a right triangle. Imagine moving that triangle to the other side so that you get a rectangle, like in the examples above. Find the area of the rectangle, thereby finding the area of the original parallelogram.

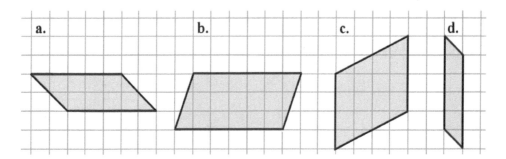

a. _____ square units   b. _____ square units

c. _____ square units   d. _____ square units

One side of the parallelogram is called the **base**. You can choose any of the four sides to be the base, but people often use the "bottom" side.

A line segment that is *perpendicular* to the base and goes from the base to the opposite side of the parallelogram is called the **altitude**.

When we do the trick of "moving the triangle," we get a rectangle. One of its sides is congruent (has the same length) to the parallelogram's *altitude*. The other side is congruent to the parallelogram's *base*.

That is why you can simply multiply **BASE × ALTITUDE** to get the area of a parallelogram.

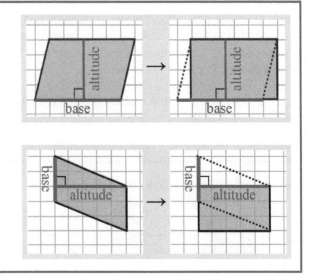

3. Draw an altitude to each parallelogram. Highlight or "thicken" the base. Then find the areas.

a. _____ sq. units

b. _____ sq. units

c. _____ sq. units

d. _____ sq. units

e. _____ sq. units

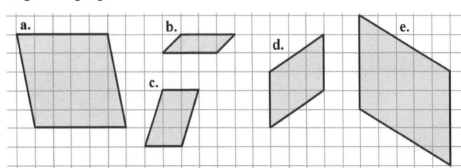

4. Find the area of the parallelogram in square *centimeters*.

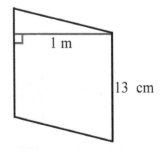

5. Find the area of the parallelogram in square meters.

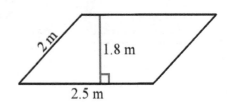

113

Sometimes the parallelogram is so "slanted" that its altitude does not even reach the base! Instead, the altitude ends at the continuation of the base (marked with a dashed line).

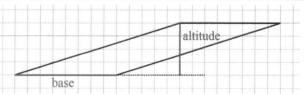

That is no problem—what matters is that the altitude is *perpendicular* to the base. We can still gets the area by multiplying the base and altitude.

In this case, the base is 8 units and the altitude is 4 units, so the area is 32 square units.

6. **a.** Draw the altitudes to the parallelograms, and mark their bases. One altitude is already done for you.

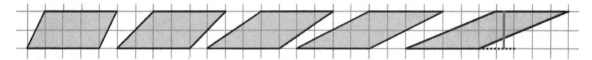

b. Find the areas of the parallelograms. What do you notice?

7. Draw at least five differently-shaped parallelograms that all have an area of 12 square units.

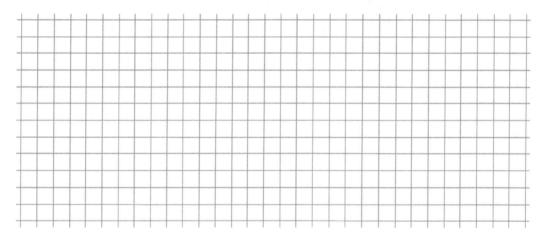

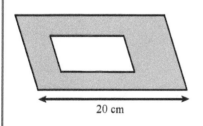

The altitude of the larger parallelogram is half of its base, and the base and altitude of the smaller parallelogram are half the base and altitude of the larger one.

Find the area of the shaded area.

Puzzle Corner

Area of Triangles

We can always put any triangle together with a copy of itself to make a parallelogram.

Therefore, **the area of the triangle must be exactly half of the area of that parallelogram.**

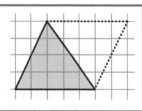

1. Find the area of the shaded triangle in the picture above.

2. Draw the corresponding parallelograms for these triangles, and find the areas of the triangles.
   Hint: Draw a line that is congruent to the base of the triangle, starting at the top vertex.

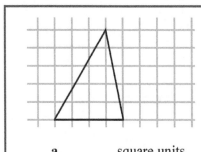

a. _____ square units

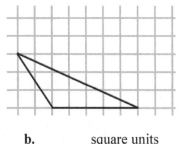

b. _____ square units

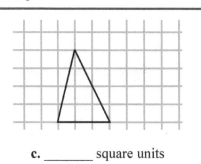

c. _____ square units

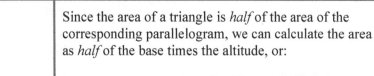

Again, we use a base and an altitude. The **base** can be any side of the triangle, though people often use the "bottom" side.

The **altitude** is *perpendicular* to the base, and it goes from the opposite vertex to the base (or to the continuation of the base).

Since the area of a triangle is *half* of the area of the corresponding parallelogram, we can calculate the area as *half* of the base times the altitude, or:

$$\text{AREA} = \frac{\text{BASE} \cdot \text{ALTITUDE}}{2}$$

You can choose any side to be the base. Here, it makes sense to choose the vertical side as the base.

The area is $\dfrac{4 \cdot 6}{2} = 12$ square units.

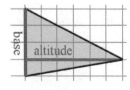

3. Draw an altitude in each triangle, and mark the base. Find the area of each triangle.

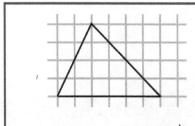

a. _____ square units

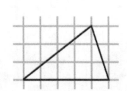

b. _____ square units

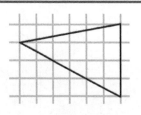

c. _____ square units

Example 1. The altitude of a triangle may fall *outside* of the triangle itself. It is still perpendicular to the base, and starts at a vertex.

The corresponding parallelogram is seen if you follow the dotted lines.

The area is $\dfrac{3 \cdot 3}{2}$ = 4 ½ square units.

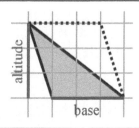

Example 2. Here it is easiest to think of the base being on the top. Again, the altitude falls outside the actual triangle.

The area is $\dfrac{5 \cdot 3}{2}$ = 7 ½ square units.

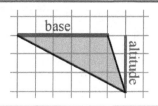

4. Draw the base and the altitude to this triangle, and find its area.

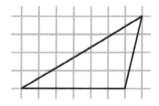

5. Draw altitudes and bases in the triangles, and find their areas.

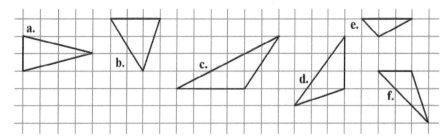

a. _____ square units **b.** _____ square units **c.** _____ square units

d. _____ square units **e.** _____ square units **f.** _____ square units

6. This figure is called a _____.

   Find its area (thinking of it being divided into three parts).

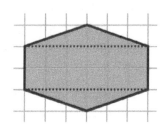

7. Draw as many different-shaped triangles as you can that have an area of 12 square units.

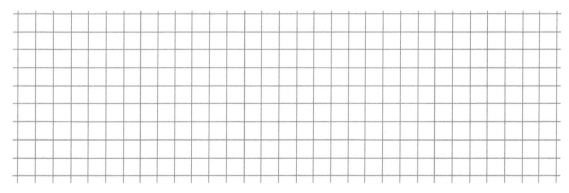

Polygons in the Coordinate Grid

Here is a neat way to **find the area of any polygon whose vertices are points in a grid.**

(1) Draw a rectangle around the polygon.

(2) Divide the area between the polygon and the rectangle into triangles and rectangles.

(3) Calculate those areas.

(4) <u>Subtract</u> the calculated areas from the total area of the large rectangle to find the area of the polygon.

Example. To find the area of the colored triangle, we draw a rectangle around it that is 3 units by 6 units. Then we find the areas marked with 1, 2, 3, 4, and 5:

1: a triangle; 3 · 3 ÷ 2 = 4.5 square units
2: a triangle; 1 · 3 ÷ 2 = 1.5 square units
3: a rectangle; 1 · 3 = 3 square units
4: a triangle; 1 · 3 ÷ 2 = 1.5 square units
5: a triangle; 1 · 3 ÷ 2 = 1.5 square units
The total for the shapes 1, 2, 3, 4, and 5 is 12 square units.

Therefore, the area of the colored triangle is 18 square units − 12 square units = 6 square units.

1. Find the areas of the shaded figures.

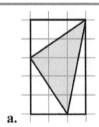

a.

b.

2. This figure is called a _____.

   Calculate its area using the three triangles.
   For each triangle, use the *vertical* side as the base.

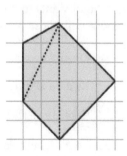

3. Let's use another way of calculating the area of the same figure.

   1. Calculate the area of the rectangle that encloses the figure.
   2. Calculate the areas of the four shaded triangles.
   3. Subtract.

   Verify that you get the same result as in exercise #1.

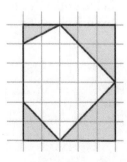

4. Find the area of the triangle with the vertices at (−8, 7), (−5, 3), and (4, 0).

5. Draw a quadrilateral in the grid with vertices (8, 5), (3, 4), (4, −5), and (7, −6).

   Then find its area.

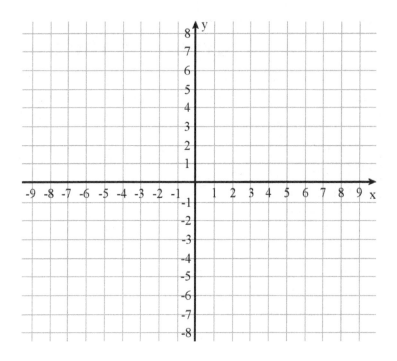

6. These are the coordinates of the foundation for a high-rise building in a special polygonal shape:

   (20, 20), (40, 40), (20, 60), (40, 60), (60, 40), (40, 20)

   a. What is the polygon called?

   b. Find its area.

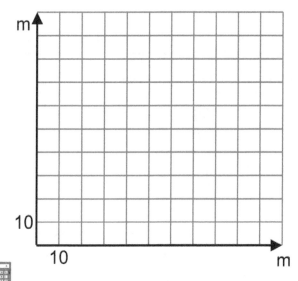

7. A hotel wants to build a rectangular swimming pool with these requirements:

   • One of its sides has to be 12.5 meters.

   • Its area should be between 100 and 140 square meters.

   • It has to be at least 10 meters away from the driveway.

   Design a rectangular pool (decide its size and location), and draw it in the grid. Give the pool's area and the distance from the pool to the driveway. Each unit in the grid is 5 m.

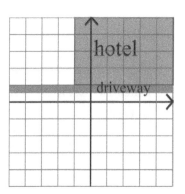

118

8. The shape of Mountain View Cemetery in Calgary, Canada, is very close to a rectangle.

   a. Use the coordinates of the points (given in meters) and calculate its area.

   b. Estimate the distance between points A and B on the map.

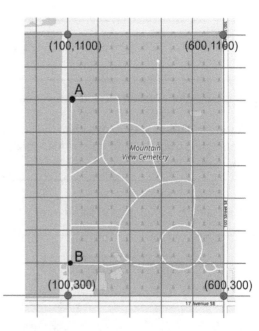

9. Find the area of a rectangle with vertices at the points (−20, 65), (−20, 30), (45, 65), and (45, 30). You can use the grid to help.

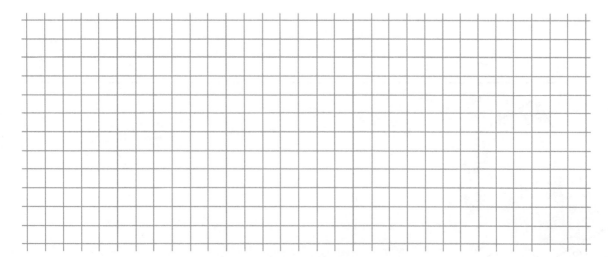

Area of Polygons

In the previous lesson we found the area of a polygon by enclosing it in a rectangle, and by using subtraction.

Another, natural way to calculate the area of a polygon is to divide the polygon into easy shapes, such as rectangles, triangles and trapezoids. Calculate the area of each shape separately, and then add them to find the total area.

1. Calculate the total area of the figures.

a.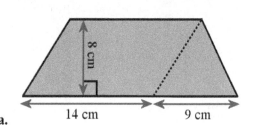

b.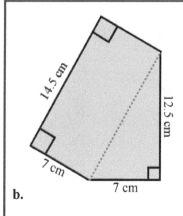

2. **a.** The side of each little square in the drawing on the right is <u>1 inch</u>. Find the area of the polygon.

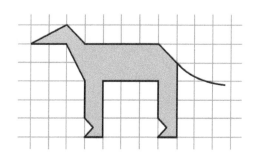

b. Imagine that the side of each little square is <u>2 inches</u> instead. What is the area now?

120

3. Find the area of the polygon.

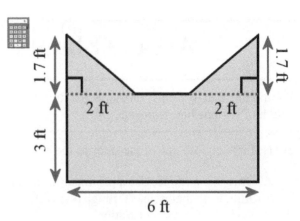

4. The vertices of a right triangle are at (17, 26), (17, −9), and (2, −9). Find its area.

5. The points (1, 2.4), (2.4, 1), (2.4, −1), (1, −2.4) are four vertices of a water fountain in the shape of a regular octagon. The other four points are found by reflecting these four in the *y*-axis.

   a. Draw the octagon.

   b. Find the length of *one* side of the fountain.

   c. Find its perimeter.

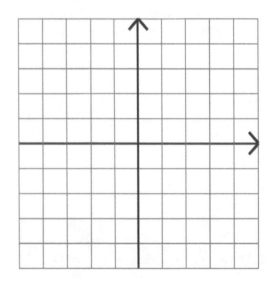

Puzzle Corner Join the following points in order with line segments. Then find the area of the resulting polygon.

(−35, −40), (−35, 40), (−20, 40), (20, −15), (20, 40), (35, 40), (35, −40), (20, −40), (−20, 15), (−20, −40) and (−35, −40)

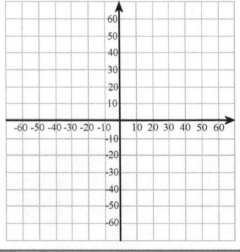

Area of Shapes Not Drawn on Grid

To find the area of a polygon that is not draw on a grid, we often need to draw altitudes and measure the length of various line segments.

1. First, choose one of the sides as the base. It can be any side!

2. Draw the altitude. Use a protractor or a triangular ruler to draw the altitude so that it goes through one vertex and is perpendicular to the base. See the illustration.

   Line up the 90°-mark on the protractor with the base of the triangle and slide it until the line you draw will pass through the vertex.

3. Measure the lengths of the altitude and base as precisely as you can with a ruler.

4. Calculate the area.

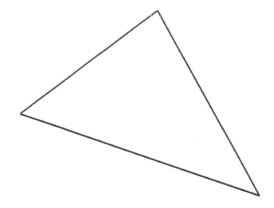

1. Find the area of this triangle in square centimeters. Round your final answer to the nearest whole square centimeter.

2. Draw your own triangle here, and find its area!

3. Find the of area of each parallelogram below, measuring the necessary parts to the nearest millimeter. Round your final answer to whole square centimeters.

   You will need to draw an altitude to each parallelogram. Use a protractor or a triangular ruler—do not "eyeball" it. The picture on the right shows how to position a protractor for drawing the altitude. Notice that the 90° mark is aligned with the base.

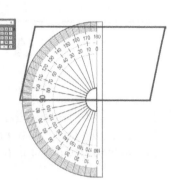

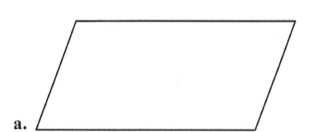
a.

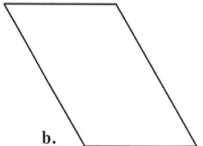
b.

4. Divide this quadrilateral into two triangles, and then find its area in square centimeters. You may use a calculator.

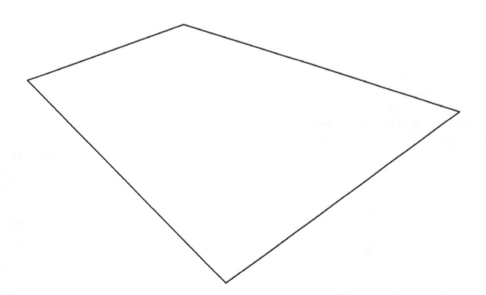

Draw a triangle with an *area* of 18 square centimeters.

Is it only possible to draw just *one* triangle with that area, or is it possible to draw several, with varying shapes/sizes?

Area and Perimeter Problems

You can use a calculator in all the problems in this lesson.

1. On paper, Mr. Smith's house plan measures 9″ × 12″.
   In reality, the house is 40 times as big. Find the
   perimeter of the house in feet and its area in square feet.

2. Find the floor area of this house.

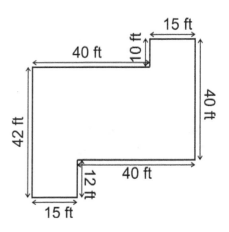

3. The picture shows a rectangular plot of land with a house in it.
   The fence around the yard has a 3 meter-wide gate.

   a. What fraction of the total area of the plot
   does the house take up?

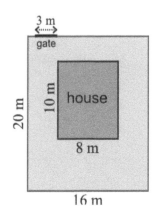

   b. Find the cost of fencing this plot when
   the fence costs $11.55 per meter, and
   the gate costs $120.

4. A father is leaving a plot of land to his two children. In his will, he specifies it to be divided into two parts as shown. The two children are not sure which part is bigger.

   a. Find the area of part 2.

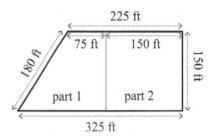

   b. Find the area of part 1.

   c. Find the area of the whole plot.

   d. Part 1 is what percentage of the total area (to the nearest percent)? And part 2?

   e. If the whole plot is valued at $45,000, find the value of part 1 and part 2 to the nearest dollar.

5. These are designs for decorative tilings. What fractional part of each design consists of the darker color?

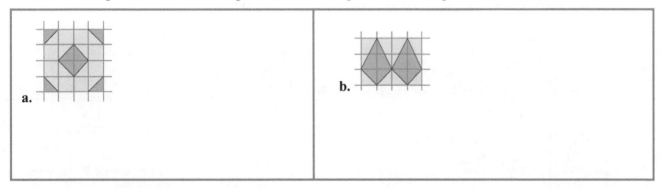

6. This is a front side of a house. Find its area. The dimensions are given in feet.

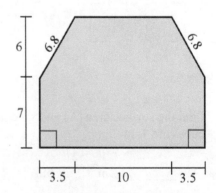

Nets and Surface Area 1

This picture shows a flat figure, called a **net**, that can be folded up to form a solid, in this case a cube.

Each face of a cube is a square. If we find the total area of its faces, we will have found the **surface area** of the cube.

Let's say that each edge of this cube measures 2 cm. Then one face would have an area of 2 cm · 2 cm = 4 cm², and the total surface area of the six faces of the cube would be 6 · 4 cm² = 24 cm².

What is its volume? Remember, **volume** has to do with how much space a figure takes up, and not with "flat" area. Volume is measured in *cubic* units, whereas area is measured in *square* units. The volume of this cube is 2 cm · 2 cm · 2 cm = (2 cm)³ = 8 cm³.

1. Which of these patterns are nets of a cube? In other words, which ones can be folded into a cube? You can copy the patterns on paper, cut them out and fold them.

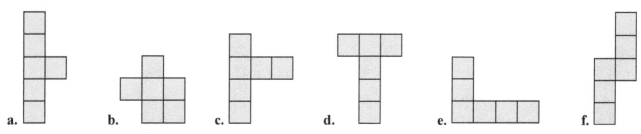

2. Match each rectangular prism (a), (b), (c) and (d) with the correct net (1), (2), (3) and (4). Again, if you would like, you can copy the nets onto paper, cut them out, and fold them.

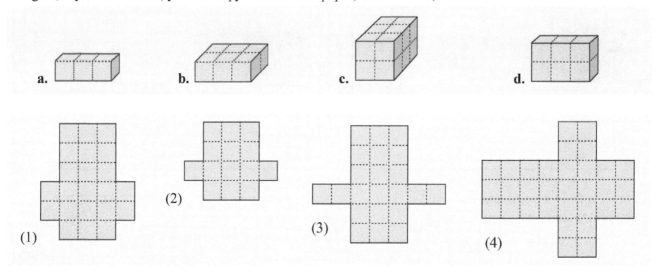

3. Find the surface area (A) and volume (V) of each rectangular prism in problem #2 if the edges of the little cubes are 1 cm long.

a. A = _____ cm²   **b.** A = _____ cm²   **c.** A = _____ cm²   **d.** A = _____ cm²

V = _____ cm³   V = _____ cm³   V = _____ cm³   V = _____ cm³

A **prism** has two identical polygons as its top and bottom faces. These polygons are called the *bases* of the prism. The bases are connected with faces that are parallelograms (and often rectangles).

Prisms are named <u>after the polygon used as the bases</u>.

A rectangular prism.
The bases are rectangles.

Two **triangular prisms**.
One is lying down, where the base is facing you.
The other is "standing up".

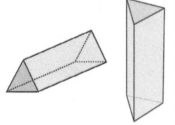

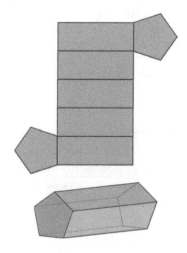

A pentagonal prism and its net. The bases are pentagons. Again, the base is not "on the bottom" but facing you.

A **pyramid** has a polygon as its bottom face (the base), and triangles as other faces.

Pyramids are named after the polygon at the base: a triangular pyramid, square pyramid, rectangular pyramid, pentagonal pyramid, and so on.

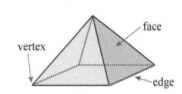

A **square pyramid** and its net.

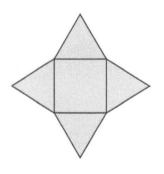

See interactive solids and their nets at the link below:
https://www.mathsisfun.com/geometry/polyhedron-models.html

4. Name the solid that can be built from each net.

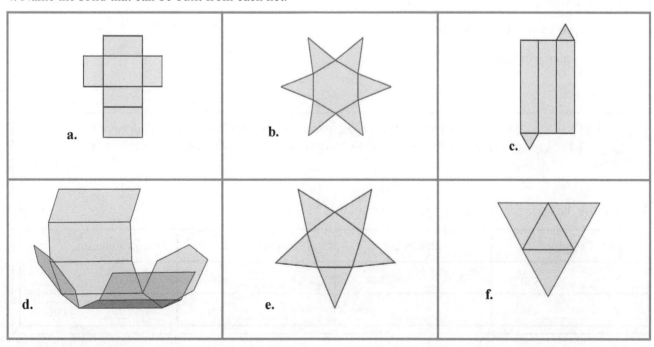

127

5. Which expression, (1), (2), or (3), can be used to calculate the surface area of this prism correctly? (You do *not* have to actually calculate the surface area.)

   1. $2 \cdot 35 \text{ cm}^2 + 2 \cdot 63 \text{ cm}^2 + 2 \cdot 45 \text{ cm}^2$

   2. $5 \text{ cm} \cdot 9 \text{ cm} \cdot 7 \text{ cm}$

   3. $5 \text{ cm} \cdot 7 \text{ cm} + 9 \text{ cm} \cdot 7 \text{ cm} + 9 \text{ cm} \cdot 5 \text{ cm}$

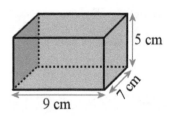

6. Ryan organized the calculation of the surface area of this prism into three parts. Write down the intermediate calculations, and solve. This way, your teacher (or others) can follow your work. Remember also to <u>include the units</u> (cm or cm²)!

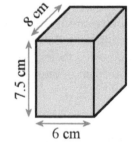

Top and bottom:

Back and front:

The two sides:

Total:

7. The surface area of a cube is 96 square inches.

   a. What is the area of one face of the cube?

   b. How long is each edge of the cube?

   c. Find the volume of the cube.

Puzzle Corner

Consider the rectangular prisms in problem #2. If the edges of the little cubes were double as long, how would that affect the surface area? Volume?

You can use the table below to investigate the situation.

Prism a.	Prism b.	Prism c.	Prism d.
A = _____ cm²	A = _____ cm²	A = _____ cm²	A = _____ cm²
V = _____ cm³	V = _____ cm³	V = _____ cm³	V = _____ cm³

Nets and Surface Area 2

You can use a calculator in all problems of this lesson.

1. Name the solid that can be built from this net, and then calculate the surface area. Write down the intermediate calculations so that your teacher (or others) can follow your work.

   solid: _____

   area of the bottom face:

   area of one of the triangles:

   total surface area:

2. Name the solid, draw a net for it, and then calculate the surface area. Write down the intermediate calculations so that your teacher (or others) can follow your work.

   solid: _____

3. A gift box is in the shape of a cube with 20-cm sides. Calculate its surface area.

4. Draw a net for each of these rectangular prisms.

a.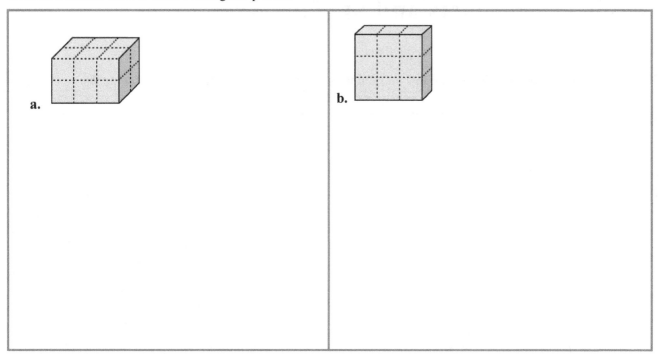

b.

5. This box of chocolate is in the shape of a triangular prism.

   a. Draw a net for the prism.

   b. Calculate its surface area.

Problems to Solve - Surface Area

You may use a calculator in all problems of this lesson.

1. The *volume* of a cube is 27 cubic meters.
   Find its surface area.

2. Name the solid that can be built from the net, and calculate the surface area. Write down the intermediate calculations so that your teacher (or others) can follow your work.

   solid: _____

   [Net showing: top rectangle 14 in. × 7 in.; middle row with 7 in. × 11 in., 14 in. × 11 in., and 7 in. × 11 in.; bottom rectangle 14 in. × 7 in.]

3. Name the solid, draw a net for it, and then calculate the surface area. Write down the intermediate calculations so that your teacher (or others) can follow your work.

   Solid: _____

   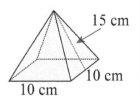

   15 cm, 10 cm, 10 cm

131

4. Annie wants to cover the sides of this box with one long piece of pretty wrapping paper. She does NOT want to cover the bottom or the top.

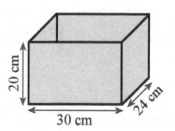

   a. What is the length and width of the piece of wrapping paper she needs?

   b. What is the area of the wrapping paper?

5. A swimming pool is in the shape of a rectangular prism.
   It is 12.5 m long, 6 m wide, and 2 m deep.

   a. Sketch the swimming pool and mark the dimensions on the sketch.

   b. In order to tile the pool, find the surface area of the pool's bottom and sides (not including the top, since it won't be tiled).

   c. Tile costs $9.90 per square meter.
      Calculate the cost of tiling the pool.

Converting Between Area Units

I could tell you conversion factors for different units of area, but instead I want to show you *how YOU can figure them out yourself!* It is actually pretty simple!

The image shows 1 square centimeter. It is divided into square millimeters. The side of the whole square is 1 cm and at the same time 10 mm.

How many square millimeters are in one square centimeter?

$1 \text{ cm}^2 = 10 \text{ mm} \cdot 10 \text{ mm} =$ _____ mm^2

1. The sides of the large square measure 1 yard. How many square feet are there in 1 square yard?

   1 sq. yd. = _____ ft × _____ ft = _____ sq. ft.

2. Imagine or sketch a square with 2-meter sides.

   The area in square *meters* is ____ m · _____ m = _____ m^2.

   The area in square *centimeters* is _____ cm · _____ cm = _____ cm^2.

3. Use similar reasoning to determine:

   a. How many square meters are in one square kilometer?

   b. How many square millimeters are in one square centimeter?

4. Measure what you need from the figures and find their areas in square millimeters and in square centimeters.

   a. _____ mm^2 b. _____ mm^2 c. _____ mm^2

   _____ cm^2 _____ cm^2 _____ cm^2

5. Think of the relationship between square centimeters and square millimeters, for example, based on the previous exercise. How can you convert 58 square centimeters into square millimeters?

6. **a.** Find the area of a 0.8 mi. by 2 mi. rectangle in square miles.

   $\boxed{1 \text{ mi} = 5{,}280 \text{ ft}}$

   b. Now find the same area in square feet.

7. A village lies within a rectangle that has 0.2 km and 0.15 km sides. Find its area in square meters.

8. Connect the dots in the figure below to get a quadrilateral, or draw your own non-rectangular quadrilateral on blank paper.

   a. Draw and measure what you need, and find the area of the quadrilateral in <u>square millimeters</u>. Round it to the nearest hundred.

   b. Calculate the area to the nearest <u>square centimeter</u>.

Volume of a Rectangular Prism with Sides of Fractional Length

Example 1. Let's imagine that the edges of this little cube each measure ½ m.

If we stack eight of them so that we get a bigger cube… we get this:

The bigger cube has <u>1 m edges, so its volume is 1 cubic meter</u>.

If eight identical little cubes make up this bigger cube, and its volume is 1 cubic meter, then <u>the volume of *one* little cube is 1/8 cubic meter</u>.

Notice: this is the same result that we get if we multiply the height, width and depth of the little cube:

$$\frac{1}{2} \text{ m} \cdot \frac{1}{2} \text{ m} \cdot \frac{1}{2} \text{ m} = \frac{1}{8} \text{ m}^3$$

1. The edges of each little cube measure ½ m. What is the total volume, in cubic meters, of these figures?

a.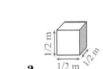

width = __1/2__ m

height = _____ m

depth = _____ m

__1__ little cube, 1/8 m³

V = _____ m³

b.

width = _____ m

height = _____ m

depth = _____ m

__8__ little cubes, each 1/8 m³

V = _____ m³

c.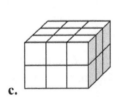

width = _____ m

height = _____ m

depth = _____ m

_____ little cubes, each 1/8 m³

V = _____ m³

d.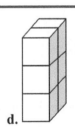

width = _____ m

height = _____ m

depth = _____ m

_____ little cubes, each 1/8 m³

V = _____ m³

2. Write a multiplication (width · depth · height) to calculate the volume of the figures (c) and (d) above, and verify that you get the same result as above.

a. V = _____ m · _____ m · _____ m

   =

b. V = _____ m · _____ m · _____ m

   =

3. Fill in.

   This time, the edges of each little cube measure 1/3 inch.

   (Mark the dimensions on the little cube.)

   We put _____ of the little cubes together to form one cubic inch (on the right). →

   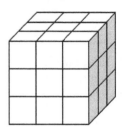

   Since the big cube measures 1 cubic inch, and there are _____ little cubes,

   the volume of each little cube is ▭/▭ cubic units. This is the same answer that we find

   by multiplying: V = ▭/▭ in · ▭/▭ in · ▭/▭ in = ▭/▭ in³.

4. Show that the volume of a box that measures 1 1/3 in. by 2 in. by 2/3 in. is indeed

   $$V = \frac{4}{3} \text{ in} \times 2 \text{ in} \times \frac{2}{3} \text{ in} = \frac{16}{9} \text{ in}^3.$$

   How?

   (i) Build a physical model or draw a sketch of the box,
   using 1/3 in. by 1/3 in. by 1/3 in. little cubes.

   (ii) Count the number of little cubes needed.

   (iii) Multiply the number of little cubes by the volume of ONE little cube.

5. Show that the volume of a box with dimensions of 3/4 in. by 2 1/4 in. by 1 in. is indeed

   $$V = \frac{3}{4} \text{ in} \times \frac{9}{4} \text{ in} \times 1 \text{ in} = \frac{27}{16} \text{ in}^3.$$

   How?

   (i) Build a physical model or draw a sketch of the box,
   using 1/4 in. by 1/4 in. by 1/4 in. little cubes.

   (ii) Count the number of little cubes needed.

   (iii) Multiply the number of little cubes by the volume of ONE little cube.

6. The edges of each little cube measure 1/3 in. What is the total volume of these figures, in cubic units?

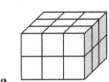

a.

width = __1__ in

height = __2/3__ in

depth = _____ in

_____ little cubes, each 1/27 in³

V = _____

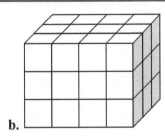

b.

width = _____ in

height = _____ in

depth = _____ in

_____ little cubes, each 1/27 in³

V = _____

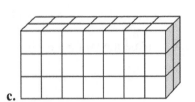

c.

width = _____ in

height = _____ in

depth = _____ in

_____ little cubes, each 1/27 in³

V = _____

You have already learned that the **volume of a rectangular prism** can be calculated by multiplying the width, depth, and height. As a formula: $V = w \times d \times h$ or just $V = wdh$.

This formula also applies when the dimensions are fractions or decimals.

Example 2. Calculate the volume of this box.

We simply multiply the three dimensions, using mixed numbers:

$\frac{7}{3}$ ft $\times \frac{3}{2}$ ft $\times 1$ ft $= \frac{7}{2}$ ft³ $= 3\frac{1}{2}$ ft³.

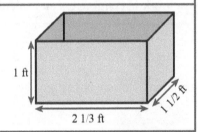

7. Write a multiplication to calculate the volume of the figures in exercise 6, and verify you get the same result.

a. V = _____ in × _____ in × _____ in =

b. V = _____ in × _____ in × _____ in =

c. V = _____ in × _____ in × _____ in =

Volume Problems

You can use a calculator in all the problems in this lesson.

1. **a.** Calculate the volume of this carton to the nearest cubic centimeter.

   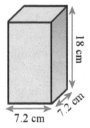

   b. Considering that 1 milliliter = 1 cubic centimeter, what is the volume of the carton in milliliters?

   c. The carton is 96% full of juice. How many milliliters of juice does it contain?

2. A company builds cube-shaped storage boxes where the edge of a cube measures half a meter.

   a. Work in meters, and using *fractions*. Calculate the volume of one such box in cubic meters.

   b. Now work in centimeters and calculate the volume of one such box in cubic centimeters.

   c. How many storage boxes are needed to have a total volume of one cubic meter?

3. Alex built a wooden crate (in the shape of a rectangular prism) measuring 0.4 m by 0.9 m, and 0.5 m high. He will fill it with dirt and plant flowers in it. In the local garden center, 40-liter bags of garden soil cost $34.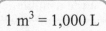

   a. If Alex only used soil from the bags, how many bags would he need to buy to fill his crate?

   b. How much would those cost him?

4. Your friend in the United States told you that his aquarium is 20 gallons. Which of the aquariums listed below is closest to the size of your friend's aquarium?

   1 L = 1,000 ml
   1,000 ml = 1,000 cm^3
   1 gallon = 3.78541 L

   Also... can you think of a way to solve this without "brute force" calculations?

	DIMENSIONS (cm)	VOLUME (L)	VOLUME (gal)
Aquarium 1	41 · 20 · 25.5		
Aquarium 2	51 · 25 · 30		
Aquarium 3	61 · 30.5 · 40.5		
Aquarium 4	91.5 · 30.5 · 43		
Aquarium 5	92.5 · 46 · 41		

5. Design a swimming pool in the shape of a rectangular prism so that:
   - It is at least 12 meters long.
   - The water will be 1.2 m deep.
   - It holds at most 100,000 liters of water.

   In other words, you need to decide its width and length. <u>Note:</u> One cubic meter is equal to 1,000 liters.

Puzzle Corner

Each edge of a "magic" cube toy measures 2¼ inches.

a. Calculate its volume.

b. How many of these cubes could you pack into a box that measures 2 ft by 1 ft by 1 ft?
Hint: Draw a sketch, and think in inches, not feet.

Chapter 9 Mixed Review

1. A family put 1/3 of 60 pounds of flour into the cellar.
   Then, they gave 3/8 of the remaining flour to a neighbor.
   How much flour did the neighbor get?
   (Lessons in Problem Solving/Ch.1)

2. **Multiply.** (Review: Multiply Decimals by Decimals/Ch.3)

a. $3 \cdot 0.3 \cdot 0.08 =$ _____	b. $7 \cdot 0.2 \cdot 1.1 =$ _____	c. $0.25 \cdot 10^5 =$ _____
d. $0.0009 \cdot 8 =$ _____	e. $0.002 \cdot 100 =$ _____	f. $3,000 \cdot 0.0007 =$ _____

3. Order the fractions from the smallest to the biggest. (Review: Add and Subtract Fractions and Mixed Numbers/Ch7)

a. $\frac{5}{6}, \frac{8}{10}, \frac{7}{8}, \frac{9}{10}, \frac{7}{10}$	b. $\frac{9}{8}, \frac{11}{10}, \frac{7}{6}, \frac{12}{10}, \frac{10}{8}$
____ < ____ < ____ < ____ < ____	____ < ____ < ____ < ____ < ____

4. Convert the measurements into the given units. (Convert Metric Measuring Units/Ch.3)

   a. 0.9 L = _____ dl = _____ cl = _____ ml

   b. 2,800 m = _____ km = _____ dm = _____ cm

   c. 56 g = _____ dg = _____ cg = _____ mg

5. Convert. Round your answers to 2 decimals in (a) - (d). In (e) and (f) use whole numbers.

a. 76 oz = _____ lb	c. 3.6 gal = _____ qt	e. 2.67 mi = _____ ft
b. 98 in = _____ ft	d. 0.483 lb = _____ oz	f. 5.09 ft = ____ ft _____ in

6. Calculate in your head. Give your answer as a decimal. (Add and Subtract Decimals/Ch.3)

a. $\frac{7}{1,000} + 0.3$	b. $1.2 + \frac{4}{100}$	c. $7.004 + \frac{5}{100}$

7. Two of these calculations are in error. Find them and correct them. (Add and Subtract Decimals/Ch.3)

   a. $0.8 + 0.03 = 0.11$ b. $2.24 - 0.04 = 2.2$ c. $0.007 + 0.7 = 0.077$

8. Solve the equations. (Review: Divide Decimals by Decimals/Ch.3)

a. $0.2m = 6$	b. $0.3p = 0.09$	c. $y - 1.077 = 0.08$

9. a. Draw a picture where there are 2 triangles for each 5 squares, and a total of 21 shapes. (Ratios and Rates *and* Unit Rates/Ch.4)

   b. The unit rates are:

   _____ squares for **1** triangle

   _____ triangles for **1** square

10. Add and subtract. (Add Integers: Counters/Ch.8)

a. $5 + (-8) =$ _____	b. $-11 + (-9) =$ _____	c. $2 + (-17) =$ _____	d. $2 - (-8) =$ _____
$(-5) + 8 =$ _____	$9 - 11 =$ _____	$-3 - 8 =$ _____	$-8 - (-2) =$ _____

11. A figure whose vertices are at $(-5, -3)$, $(-1, -3)$, $(0, -5)$ and $(-7, -5)$ is transformed this way:

    1. It is reflected in the *x*-axis.
    2. It is moved four units to the right, five down.
    3. It is reflected in the *y*-axis.

    Give the coordinates of its vertices after all three transformations.
    (Coordinate Grid/Ch.8)

12. Draw a triangle whose vertices are at $(-3, -4)$, $(5, -4)$ and $(2, 7)$.
    (Area of Triangles/Polygons in the Coordinate Grid/Ch.9)

    Draw an altitude to the triangle.

    Find its area.

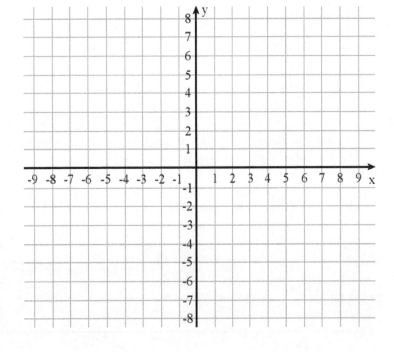

13. A mole is digging a tunnel at the speed of 4 meters per hour. (Using Two Variables/Ch.2)

   a. Choose a letter variable to represent the time the mole has dug and another to represent the length (distance) of tunnel it has dug.

   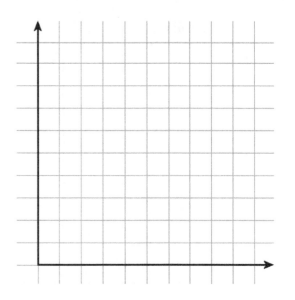

   b. Fill in the table. Plot the points.

time (hours)	0	1	2	3	4	5	6	7	8	9
distance (meters)										

   c. Write an equation relating the two variables.

   d. Which is the independent variable?

14. Fill in the blanks and give examples. (Divide Fractions: Reciprocal Number/Ch.7)

   a. Dividing a number by 5 is the same as multiplying it by ____. Example:

   b. Dividing a number by $\frac{2}{3}$ is the same as multiplying it by ____. Example:

15. Write as percentages. If necessary, round your answers to the nearest percent. (Percent/Ch.5)

   a. 5/8

   b. 6/25

16. Draw a triangle with 55° and 29° angles, and a 6-cm side between those angles. Use a notebook or blank paper. (Triangles Review/Ch.9)

17. Draw a rhombus with 7.5 cm sides, and one 66° angle. Use a notebook or blank paper. (Quadrilaterals Review/Ch.9)

Puzzle Corner Find the missing factors.

a. $\frac{1}{5} \cdot \underline{} = \frac{1}{20}$ **b.** $\frac{1}{5} \cdot \underline{} = 2$ **c.** $\frac{5}{6} \cdot \underline{} = \frac{1}{3}$

Geometry Review

You can use a calculator in all the problems in this lesson.

1. Explain how the area of the triangle is related to the area of the parallelogram.

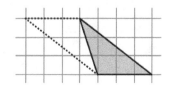

2. Find the area of the quadrilaterals in square units.

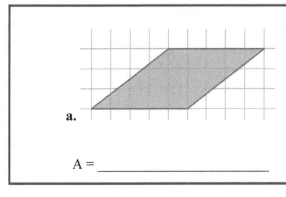

a.

A = _____

b.

A = _____

3. **a.** Jeremy planted a garden in the shape of the diagram at the right. Find the area of Jeremy's garden.

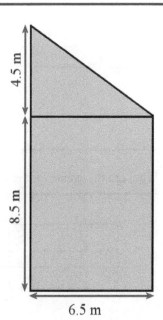

b. Jeremy planted a rectangular section measuring 3.5 m by 3 m with green beans. What percentage of his garden did he plant with green beans?

4. Find the area of this triangle:

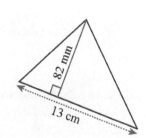

a. in square centimeters

b. in square millimeters

143

5. Draw a net and calculate the surface area. Round your answer to the nearest square centimeter.

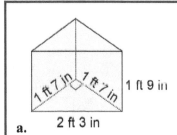
a. 2 ft 3 in

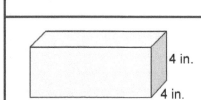
b. 11 3/4 in., 4 in., 4 in.

6. What are the names of the solids that can be constructed from these nets?

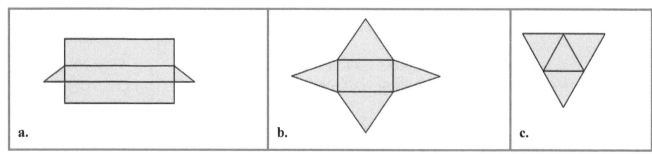

a.   b.   c.

7. What solid can you build from this net?

Calculate its surface area to the nearest square centimeter.

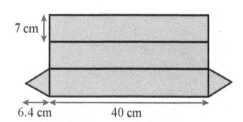
7 cm, 6.4 cm, 40 cm

8. The edges of each little cube measure 1/3 m.

   What is the total volume, in cubic meters, of the figure at the right?

   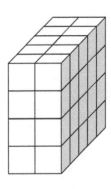

9. This building has three stories. Calculate the volume of one story.

   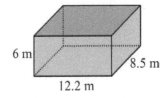

10. An aquarium measures 50 cm · 30 cm on the bottom, and its height is 40 cm. It is 4/5 filled with water.

    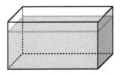

    How many cubic centimeters of water is in it?

    How many milliliters of water is in it?
    (One cubic centimeter is one milliliter.)

    How many liters?

Chapter 10: Statistics
Introduction

The fundamental theme in our study of statistics is the concept of *distribution*. In the first lesson, students learn what a distribution is—basically, it is *how* the data is distributed. The distribution can be described by its center, spread and overall shape. The shape is read from a graph, such as a dot plot or a bar graph.

Two major concepts when summarizing and analyzing distributions are its center and its variability. First we study the center, in the lessons about mean, median, and mode. Students not only learn to calculate these values, but also relate the choice of measures of center to the shape of the data distribution and the type of data.

Next, we study measures of variation, starting with range and interquartile range. Students use these measures in the following lesson, as they both read and draw boxplots.

The lesson *Mean Absolute Deviation* introduces students to this measure of variation. It takes many calculations, and the lesson gives instructions on how to calculate it using a spreadsheet program (such as Excel or LibreOffice Calc).

Next, students learn to make histograms. They will also continue summarizing distributions by describing their shape, and giving a measure of center and a measure of variability. The lesson on stem-and-leaf plots is optional.

There are some videos available for these topics at **https://www.mathmammoth.com/videos/** (choose 6th grade).

The Lessons in Chapter 10

	page	span
Understanding Distributions	149	*5 pages*
Mean, Median and Mode	154	*2 pages*
Using Mean, Median and Mode	156	*2 pages*
Range and Interquartile Range	158	*2 pages*
Boxplots	160	*3 pages*
Mean Absolute Deviation	163	*4 pages*
Making Histograms	167	*3 pages*
Summarizing Statistical Distributions	170	*4 pages*
Stem-and-Leaf-Plots	174	*2 pages*
Chapter 10 Mixed Review	176	*3 pages*
Statistics Review	179	*4 pages*

Helpful Resources on the Internet

We have compiled a list of Internet resources that match the topics in this chapter. This list of links includes web pages that offer:

- **online practice** for concepts;
- online **games**, or occasionally, printable games;
- **animations** and interactive **illustrations** of math concepts;
- **articles** that teach a math concept.

We heartily recommend you take a look at the list. Many of our customers love using these resources to supplement the bookwork. You can use the resources as you see fit for extra practice, to illustrate a concept better, and even just for some fun. Enjoy!

https://l.mathmammoth.com/gr6ch10

Understanding Distributions

A **statistical question** is a question where we expect a range of *variability* in the answers to the question.

For example, "How old am I?" is *not* a statistical question (there is only one answer), but "How old are the students in my school?" *is* a statistical question because we expect the students' ages not to be all the same.

"How much does this TV cost?" is *not* a statistical question because we expect there to be just one answer.

"How much does this TV cost in various stores around town?" *is* a statistical question, because we expect a number of different answers: the prices in different stores will vary.

To answer a statistical question we collect a set of **data** (many answers). The data can be displayed in some kind of a graph, such as a bar graph, a histogram, or a dot plot.

This is a **dot plot** showing the ages of the participants in a website-building class. Each dot in the plot signifies one observation. For example, we can see there was one 13-year old and two 14-year olds in the class.

The dot plot shows us the **distribution** of the data: it shows how many times (the frequency) each particular value (age in this case) occurs in the data.

This distribution is actually **bimodal,** or "double-peaked". This means it has two "centers": one around 11-12 years, and another around 17-18 years.

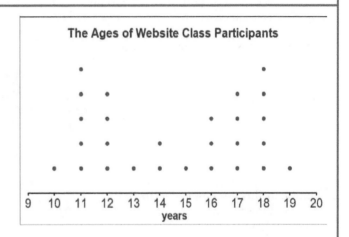

1. Are these statistical questions? If not, change the question so that it becomes a statistical question.

   a. What color are my teacher's eyes?

   b. How much money do the students in this university spend for lunch?

   c. How much money do working adults in Romania earn?

   d. How many children in the United States use a cell phone regularly?

   e. What is the minimum wage in Ohio?

   f. How many sunny days were there in August, 2020, in London?

   g. How many pets does my friend have?

149

2. Is this graph based on a statistical question?

Why or why not?

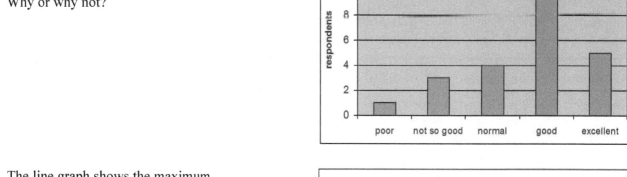

3. The line graph shows the maximum temperatures in New York for each month of a certain year.

Is this graph based on a statistical question?

Why or why not?

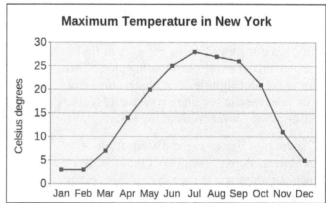

4. The title of this dot plot is not the best. But, could the plot be based on a statistical question?

If yes, give it a better, more specific, title. Imagine what situation and what question might have produced the data.

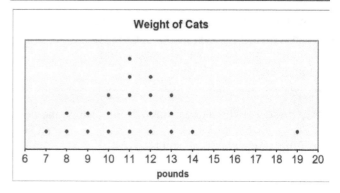

5. Change each question from a non-statistical question to a statistical question, and vice versa.

   a. What shampoo do you use?

   b. How cold was it yesterday where you live?

   c. How old are people in Germany when they marry (the first time)?

   d. How long does it take for our company's packages to reach the customers?

We are often interested in the **center**, **spread** and **overall shape** of the distribution. Those three things can summarize for us what is important about the distribution.

The **center** of a distribution has to do with where its peak is. We can use mean, median and mode to characterize the central tendency of a distribution. We will study those in detail in the next lesson.

These three dot plots show how the **spread** of a distribution can vary. This means how the data items themselves are spread—whether they are "spread" all over, or tightly concentrated near some value, or somewhat concentrated around some value. We will study more about spread in another lesson.

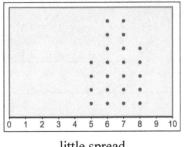
little spread

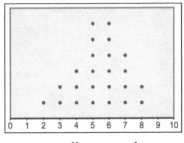
medium spread

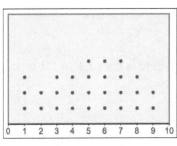
large spread

The distribution can have many varying overall **shapes.** For example:

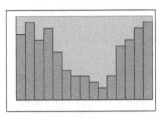
U-shaped

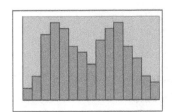
double-peaked (bimodal)

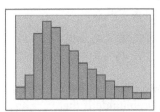
asymmetrical, right-tailed (a.k.a. right-skewed)

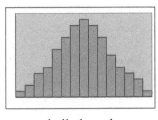
bell-shaped

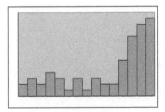
J-shaped
(can also be mirrored where most of the values are at the left)

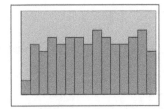
rectangular

In addition to its overall shape, a distribution may have a gap, an outlier, or a cluster:

This distribution has a **gap** from 19 to 22:	In this distribution, 9 is an **outlier** — a data item whose value is considerably less or more than all the others.	This distribution has a bell shape overall (with a peak at 18), but also a **cluster** or a smaller peak at 8-10.
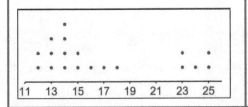	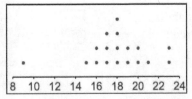	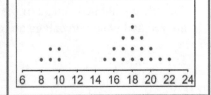

6. Anne asked her classmates the question, "How tall are you?" The histogram shows the distribution of her data.

   a. Describe the overall shape of the distribution, and also include if there are any striking deviations from the overall pattern (gaps, outliers, or clusters).

   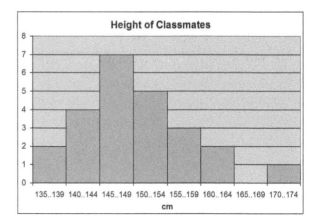

   b. Where is the peak of the distribution?

   c. How many observations are there?

7. Make a dot plot from this data (weekly work hours of a restaurant's employees). You need to place a dot for each observation.

   48 45 46 41 42 42 43 43 42 42 41
   41 45 49 40 41 41 42 46 47 42 40

   a. Describe the overall shape of the distribution. Also include if there are any striking deviations from the overall pattern (gaps, outliers, or clusters).

   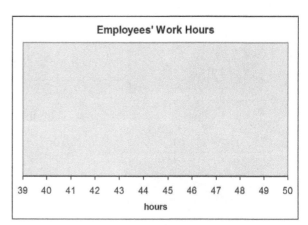

   b. Where is the peak of the distribution?

   c. How many observations are there?

8. a. Does this graph show a statistical distribution? Why or why not?

   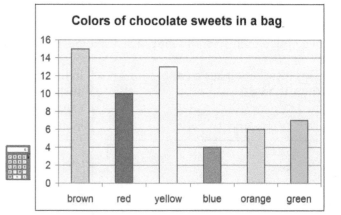

   b. Calculate what percentage of the candies are red and what percentage are green.

9. First, count the number of letters in these expressions for "Thank You" from various languages and fill in the empty column in the table. Next, label the number line below the dot plot so that all of the data will fit. Finally, plot the data.

Language	Spelling	Number of letters
Africaans	dankee	
Arabic	shukran	
Chinese, Cantanese	do jeh	
Chinese, Mandarin	xie xie	
Czech	dêkuji	
Danish	tak	
English	thank you	
Finnish	kiitos	
French	merci	
German	danke	
Greek	efharisto	
Hawaiian	mahalo	
Hebrew	toda	
Hindi	sukria	
Italian	grazie	
Japanese	arigato	
Korean	kamsa hamnida	
Norwegian	takk	
Philippines (Tagalog)	salamat po	
Polish	dziekuje	
Portuguese	obrigado	
Russian	spasibo	
Spanish	gracias	
Sri Lanka (Sinhak)	istutiy	
Swahili	asante	
Swedish	tack	
Thai	khop khun krab	
Turkish	tesekkür ederim	
Vietnamese	ca'm on	

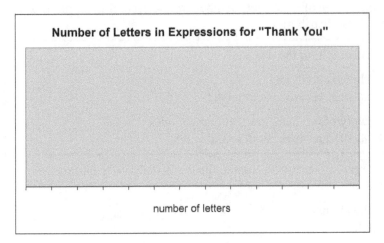

Number of Letters in Expressions for "Thank You"

number of letters

a. Describe the overall shape of the distribution. Also include if there are any striking deviations from the overall pattern (gaps, outliers, or clusters).

b. Where is the peak of the distribution?

c. How many observations are there?

Mean, Median and Mode

Mean, median and **mode** are all measures for the *center* of a data set. In other words, each of them gives us a *single number* that indicates a "middle point" of the distribution.

The **mode** is the most commonly occurring data item within the data set.

- If no item occurs more often than others, there is no mode.
  Example 1. The data set {*bear, parrot, cat, dog, lizard*} has no mode.

- If two (or three, four, *etc.*) items occur equally often, there are that many modes.
  Example 2. The data set {3, 3, 6, 6, 7, 8, 8, 10} has three modes: 3, 6 and 8.

The **median** is the *middle* item after the data is organized from the least to the greatest. Exactly half of the data is before the median, and the other half is after.

- If there is an even number of data items, the median is the average of the two items in the middle.

Example 3. Find the median of the ages of a group of children:
3, 3, 4, 6, 6, 7, 8, 8, **8**, **8**, 8, 9, 9, 9, 9, 10, 10, 10

There are 18 data items and they are already in order. The median is the "middle item", in this case the average of the 9th and 10th ages, which are both 8. So the median is 8. It matches well with the peak in the plot of the distribution.

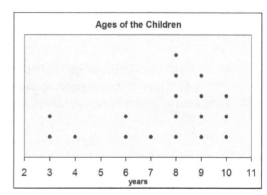

What is the mode in this example?

The **mean**, or the **average**, is calculated by adding all the data items, then dividing by the number of items.

Example 4. Mia's scores on her spelling tests were 80%, 72%, 88%, 92% and 79%. What was her average score?

We calculate the mean by adding the five scores and dividing by 5: $\dfrac{80 + 72 + 88 + 92 + 79}{5} = 82.2\%$

1. Find the median and mode of these data sets.

   a. 20, 25, 21, 30, 29, 24, 18, 32, 25, 26, 25 (ages of participants in a parenting class)

   median _____   mode _____

   b. 1, 1, 0, 2, 2, 2, 3, 1, 2, 2, 1 (the number of cars per household, for 11 households on Meadow Street)

   median _____   mode _____

   c. 80, 85, 80, 90, 70, 75, 90, 85, 100, 80 (Alice's quiz scores in algebra class)

   median _____   mode _____

   d. sandals, crocs, tennis shoes, crocs, dress shoes, sandals (types of shoes Emma keeps on her shoe rack)

   mode _____

2. Joe practices swimming. These are the times, in seconds, it took him to swim 50 m freestyle, on six different days last week: 29.76 28.45 28.12 30.73 30.48 29.57. Find his average time.

3. Find the mean, median and mode of the data sets. Draw a dot plot.

   a. Ages of children in an art club:
   4, 8, 2, 5, 5, 9, 3, 6, 5, 4, 4, 5, 1

   mean _____   median _____

   mode _____

   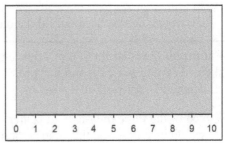

   Notice: All three measures are close to each other.
   This is not surprising, because this particular distribution
   is <u>bell-shaped</u> and has a very clear central peak.

   b. The number of sick days that a bakery's employees had last year:
   1, 1, 1, 2, 2, 2, 2, 3, 3, 4, 6, 7, 7, 8, 8, 8, 9, 9, 9, 9, 9, 10, 10

   mean _____   median _____

   mode _____

   Shape of the distribution: _____

   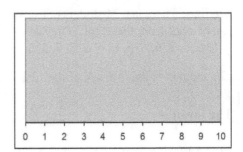

   Notice: Because of the odd shape of the distribution,
   median and mean do not describe the peaks at all.

4. These are the marks a group of students got in a course about electricity.

   a. Make a bar graph from the data.

   b. Before you go on, look at your graph and make a guess as to what the mean and median will be (approximately).

   mean _____   median _____

   c. Now find the mean, median and mode.

   mean _____   median _____   mode _____

Marks	Students
1	2
2	3
3	5
4	7
5	10

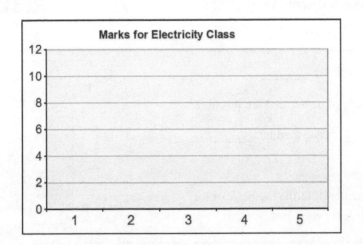

Using Mean, Median, and Mode

• The *mode* can be used with any type of data. • The *median* can only be used if the data can be put in order. • The *mean* can be used only if the data is numerical.	Whether you use mean, median, or mode depends both on • the **type of data** and • the **shape of the distribution.**

Example. This distribution of science quiz scores is heavily skewed (asymmetrical), and its "peak" is at 6. Clearly, most students did very well on the quiz.

Which of the three measures of center — mean, median, or mode — would best describe this distribution?

Mode: We can see from the graph that the mode is 6.

Median: There are 24 students. The students' actual scores can be read from the graph. They are 1, 2, 3, 3, 3, 4, 4, 5, 5, 5, 5, 5, 5, 5, 6, 6, 6, 6, 6, 6, 6, 6, 6, 6.

The median is the average of the 12th and 13th scores, which is 5.

The mean is $\dfrac{1 + 2 + 3 \cdot 3 + 2 \cdot 4 + 7 \cdot 5 + 10 \cdot 6}{24} = 4.79167 \approx 4.79$.

Notice that the mean is less than 5, but the two highest bars on the graph are at 5 and 6. In this case, the mean does *not* describe the peak of the distribution very well because it actually falls outside the peak!

The median describes the peak reasonably well, but the mode is actually the best in this situation.

1. Fill in.

a. Is the original data numerical? _____

Calculate those measures of center that are possible.

The mode: _____ The median: _____

The mean: _____

Which measure(s) of center describe the peak of the distribution well?

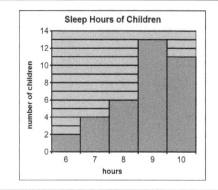

b. Is the original data numerical? _____

Calculate those measures of center that are possible.

The mode: _____ The median: _____

The mean: _____

Which measure(s) of center describe the peak of the distribution well?

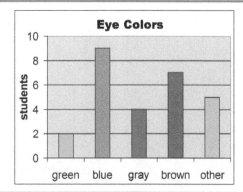

156

- Mean works best if the distribution is fairly close to a bell shape and does not have outliers. An outlier easily throws off the mean, and then the mean does not accurately describe the center of the data.
- If the distribution is very skewed or has outliers, it is better to use median than mean. Median is not sensitive to outliers — it doesn't get thrown off by an outlier (neither does the mode).

2. Judith asked 55 teenagers about how much money they spent to purchase a Mother's Day gift.

   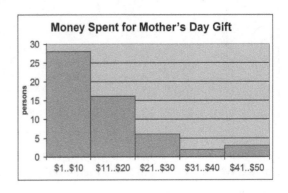

   a. The mean is $11 and the median is $9. Which of the two better describes this data? Also, explain how your choice relates to the shape of the distribution.

   b. *Approximately* what percentage of these teenagers spent $10 or less on a Mother's Day gift?

3. a. Find the mean, median, and mode of this data set: 3, 4, 4, 5, 5, 5, 5, 6, 8, 25. Note that the distribution has an outlier at 25.

   mean _____ median _____ mode _____

   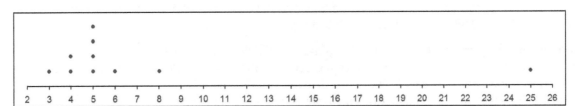

   b. Which of the three, mean, median, or mode, best describes the center of this data?

   Clearly, either the _____ or the _____, but *not* the _____!

   The _____ is off from the central peak of the distribution.

   c. Calculate the mean again if the outlier 25 is omitted. After all, it is so different from the other data items, it could even be a typing error!

4. a. Describe the overall shape or pattern of this distribution.

   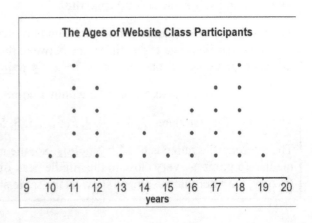

   b. Find the mode and the median.

   mode _____ median _____

   c. Which of the two better describes this distribution, and why?

Range and Interquartile Range

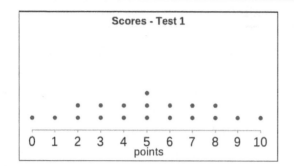
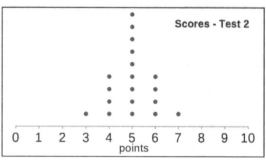

Example 1. Look at the two graphs. The first gives the scores for test 1 and the second for test 2. *Both* sets of data have a mean of 5.0 and a median of 5. Yet the distributions are very different.

How? In test 1, the students got a wide range of different scores; the data is very scattered and **varies a lot**. In test 2, nearly all of the students got a score from 4 to 6. The data is concentrated, or *clustered*, around 5.

We have several ways of measuring the variation in a distribution. One way is to use **range**. Simply put, range is **the difference between the largest and smallest data items.**

For test 1, the smallest score is 0 and the largest is 10 so the range is 10. For test 2, the smallest score is 3 and the largest is 7 so the range is 4. Clearly, the range is much smaller for test 2, indicating the data is much more concentrated than in test 1.

Another measure of variation is the **interquartile range**.

To determine this measure, we first identify the **quartiles**, which are the numbers that divide the data into quarters. The **interquartile range is the difference between the first and third quartiles**. Since the quartiles divide the data into quarters, exactly half of it lies between the first and third quartiles—and it is the middlemost half of the data. The smaller this measure is, the more concentrated the data is.

Example 2. The scores for test 1 are: 0, 1, 2, 2, 3, 3, 4, 4, 5, 5, 5, 6, 6, 7, 7, 8, 8, 9, 10. Let's now find the interquartile range. For that, we need to divide the data into quarters.

The median naturally divides the data into two halves: 0, 1, 2, 2, 3, 3, 4, 4, 5, **5**, 5, 6, 6, 7, 7, 8, 8, 9, 10.

Now we take the lower half of the data, *excluding the median*, and find *its* median: 0, 1, 2, 2, **3**, 3, 4, 4, 5. That is the **first quartile.**

Similarly, the median of the upper half of the data is the **third quartile**: 5, 6, 6, 7, **7**, 8, 8, 9, 10

The median itself is the **second quartile.**

Together, the three quartiles divide the data into quarters. The interquartile range is the difference between the third and first quartile, or in this case 7 − 3 = **4 points**.

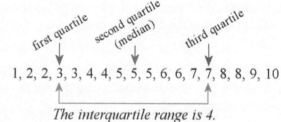

The interquartile range is 4.

So, exactly half of the test scores lie within 4 points (from 3 to 7 points) around the median of 5 points.

The scores for test 2 are: 3, 4, 4, 4, **4**, 5, 5, 5, 5, **5**, 5, 5, 5, 5, **6**, 6, 6, 6, 7. The quartiles are marked in bold.

The interquartile range is 6 − 4 = **2 points**. So, the middlemost half of the data lies within only 2 points of the median (5 points)—very close to the middle peak of the distribution. This is what we also see in the graph. Clearly, the interquartile ranges show us the same story: the data for test 1 varies much more than for test 2.

1. Find the quartiles and the interquartile range of the data sets.

   a. 5, 5, 6, 6, 7, 7, 7, 7, 7, 7, 8, 8, 8, 9, 10, 10

   first quartile _____ median _____ third quartile _____ interquartile range _____

   b. 2, 2, 3, 4, 5, 5, 5, 5, 6, 6, 6, 7, 7, 7, 9, 9

   first quartile _____ median _____ third quartile _____ interquartile range _____

   c. Let's say the data sets in (1a) and (1b) are the quiz scores of two groups (A and B) of students.

   Which group did better in general? _____ In which group did the quiz scores vary more? _____

   Make bar graphs for the quiz scores of the two groups. Note how the graphs, too, show the answers to the above questions.

2. Find the asked statistical measures of the data sets.

   a. The height of some children in centimeters:

   136 138 139 139 140 140 140 140 140 141 141 141 142 144 144 145 147

   1st quartile _____ median _____ 3rd quartile _____

   interquartile range _____ range _____

   b. The number of paid vacation days in a year of the employees in a small firm:

   6 8 10 10 11 11 12 12 12 13 13 14 14 14 15 17 18 20 24

   1st quartile _____ median _____ 3rd quartile _____

   interquartile range _____ range _____

   Use the measures you just found, and fill in:

   Half of the employees have from _____ to _____ vacation days in a year.

Boxplots

Boxplots or **box-and-whisker plots** are simple graphs on a number line that use a box with "whiskers" to visually show the quartiles of the data. Boxplots show us a **five-number summary** of the data: the minimum, the 1st quartile, the median, the 3rd quartile and the maximum.

Example 1. We have already looked at data representing the prices of hair dryers in three stores (in dollars).

14 15 19 20 20 20 21 24 25 34 34 35 35 37 42 45 55

Five-number summary:

Minimum: $14

First quartile: $20

Median: $25

Third quartile: $36

Maximum: $55

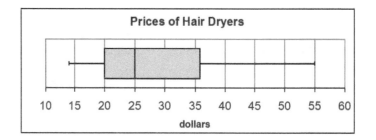

The box itself starts at the 1st quartile and ends at the 3rd quartile. Therefore, its width is the interquartile range. We draw a line in the box marking the median ($25).

The boxplot also has two "whiskers". The first whisker starts at the minimum ($14) and ends at the first quartile. The other whisker is drawn from the third quartile to the maximum ($55).

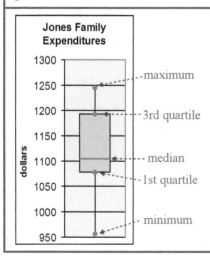

Example 2. This boxplot shows the Jones family's monthly expenditures over a 12-month period. This time the boxplot is drawn vertically.

The box extends over half the data. This means that half the time, they spend from about $1,080 to $1,190 monthly. But sometimes they spend only about $950, and sometimes up to $1,245 in a month.

Five-number summary:

Minimum: $956

First quartile: $1,078

Median: $1,105

Third quartile: $1,193

Maximum: $1,245

1. **a.** Read the five-number summary from the boxplot.

   Minimum:

   First quartile:

   Median:

   Third quartile:

   Maximum:

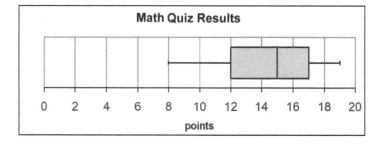

   b. Look at the box and fill in: Half the students got between _____ and _____ points in the quiz.

   c. Do you think the quiz went well?

2. Make the five-number summary for the data sets and draw a boxplot. Hint: First draw a number line.
   Note: the scaling on the number line does not have to be by ones.

   a. The number of rainy days in August, in 20 different years:

   2 3 4 4 5 6 6 6 7 7 7 8 8 8 8 9 9 11 13 16

   Minimum: _____   First quartile: _____   Median: _____   Third quartile: _____   Maximum: _____

   b. Science test scores for a 7th grade class:

   46 55 58 60 62 64 65 66 66 68 70 70 71 72 72 73 75 78 81 82 85

   Minimum: _____   First quartile: _____   Median: _____   Third quartile: _____   Maximum: _____

3. The dot plot shows the ages of a certain group of children. Make a corresponding boxplot from this data. Plot it right under the dot plot for an easy comparison.

   Hint: first write the data items as a list, using the dot plot.

   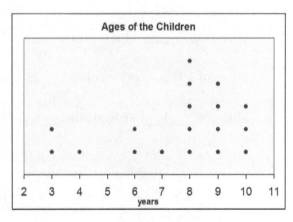

   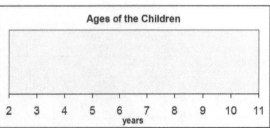

> Boxplots provide an easy way to visually compare two or more data sets, since we can easily see the middle points (medians) and the variability of the data. Note:
> - If a part of the box or a whisker is "short", the data in that part is concentrated compared to other parts.
> - If a part of a box or a whisker is long, the data in that part is scattered (spread out a lot).

4. Scientists gave three different groups of children some memory tests. The box plot shows the results.

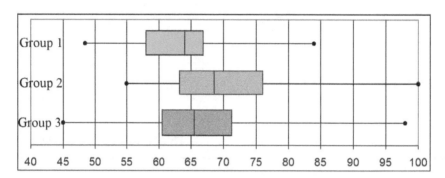

   a. Which group did the best overall?

   How do you know?

   b. Which group varied the most overall in their results?   Which group varied the least?

   How do you know?

5. **a.** Draw two side-by-side boxplots to compare the prices of various 1,000-piece 3-D and 2-D puzzles.

   b. Do the prices of the 3-D or 2-D puzzles vary more?

   c. Which type of puzzles are cheaper overall?

   prices of 2-D puzzles (in dollars):   12 14 15 15 16 16 18 20 21 22 24 27 29 35 40

   Minimum: _____ First quartile: _____ Median: _____ Third quartile: _____ Maximum: _____

   prices of 3-D puzzles (in dollars):   23 26 27 29 29 29 30 30 30 30 31 31 33 36 38

   Minimum: _____ First quartile: _____ Median: _____ Third quartile: _____ Maximum: _____

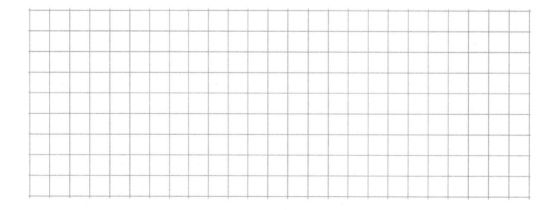

Mean Absolute Deviation

Mean absolute deviation is another measure of variation. Use it <u>only if</u> mean is the chosen measure of center.

To be brief, mean absolute deviation measures *how much, on average, the various data items deviate (differ) from the mean.* In other words, for each data item, we calculate how much it differs from the mean, and then we calculate the average of those differences.

It's called "absolute" deviation because we use the absolute values of the differences from the mean. In other words, those differences are always taken to be positive, never negative.

Example 1. Calculate the mean and the mean absolute deviation for the ages of people in a gymnastics group:

62 60 65 63 70 64 78 71 68 66 70

The mean is (62 + 60 + 65 + 63 + 70 + 64 + 78 + 71 + 68 + 66 + 70) ÷ 11 = 67 years.

So the average age of the members is 67 years. But on average, how much do their ages *differ* from this mean of 67 years? That is what mean absolute deviation will tell us.

The table shows how the calculations can be arranged. We put the data items in one column, and then we calculate the difference of each data item from the mean in another column. You can skip the column titled "difference from the mean", and go direct to the absolute difference, if you like.

The mean absolute deviation is abbreviated as *m.a.d.* in the bottom row. It is calculated as the mean of the numbers in the last column (the absolute differences), and the answer is 4 years.

What does it mean to say that the mean absolute deviation is 4? It means that, on average, the member's ages differ from the mean of 67 years by 4 years.

	age	difference from the mean	absolute difference
	62	-5	5
	60	-7	7
	65	-2	2
	63	-4	4
	70	3	3
	64	-3	3
	78	11	11
	71	4	4
	68	1	1
	66	-1	1
	70	3	3
mean	67	*m.a.d.*	4

Example 2. Calculate the mean and the mean absolute deviation for the data set 3, 4, 4, 5, 5, 5, 5, 6, 8.

The mean is (3 + 4 + 4 + 4 · 5 + 6 + 8) ÷ 9 = 45 ÷ 9 = 5. This time, let's arrange our calculation for the m.a.d. in a table horizontally.

data item	3	4	4	5	5	5	5	6	8
abs. difference from mean	2	1	1	0	0	0	0	1	3

The first row lists the data items. The second row lists the *absolute* difference of each data item from the mean. This means we took each difference as *positive*.

Lastly, calculate the mean of the absolute differences: (2 + 1 + 1 + 0 + 0 + 0 + 0 + 1 + 3) ÷ 9 = 8 ÷ 9 = 8/9 or about 0.89.

This time each data item differs from the mean of 5 about 0.89 units, on average.

Calculating mean absolute deviation using a spreadsheet program (Excel, LibreOffice Calc, etc.)

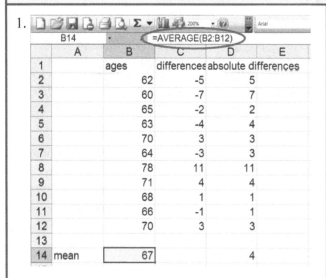

1. To calculate the mean of a set of data, in the cell where you want the calculation to appear, type:

=AVERAGE(B2:B12)

When you type the formula in the cell, it appears in the formula bar at the top, as in the image. A formula always starts with an equals sign.

Press "ENTER" to see the answer, 67.

2. Next we calculate the difference between each item of data and the mean.

Type "=B2 - B14" to subtract the values in cells B2 and B14.

The dollar signs in **B14** make it an **absolute reference**, so it doesn't change when you copy and paste the formula into another cell. Pasting the formula into the cells below is a quick way to get the spreadsheet to calculate those values, too.

3. Now we calculate the absolute value of each difference.

In cell D2 type "=ABS(C2)" to calculate the absolute value of the number in cell C2. Then copy cell D2 and paste it into the cells below it to copy the formula and adjust the reference in it automatically.

4. Lastly, we are ready to calculate the mean absolute deviation by taking the average of the values in cells D2 to D12. In the cell where you want the value to appear, type "=AVERAGE(D2:D12)".

The answer "4" will then appear in the cell after you press "ENTER."

1. Calculate the mean absolute deviation for the data sets, to two decimals. You can use the tables to help you.

   a. 6, 6, 8, 8, 8, 8, 9, 9, 9, 10, 10, 10, 11, 14 (ages of a group of children)

age															mean	
abs. difference from mean															m.a.d.	

   b. 80, 85, 80, 90, 70, 75, 90, 85, 100, 80 (Alice's quiz scores in algebra class)

points											mean	
abs. difference from mean											m.a.d.	

   c. 6, 3, 4, 12, 8, 8, 2, 6, 5 (The number of times a group of teenage girls went to the hair salon last year)

times										mean	
abs. difference from mean										m.a.d.	

2. **a.** Calculate the mean and the mean absolute deviation for this data, to three decimal digits.

Prices of MP3 players (dollars)														
price	29	30	33	34	34	35	35	35	36	36	37	39	42	mean	
abs. difference from mean														m.a.d.	

   b. Draw a dot plot for the data, and fill in the details.

   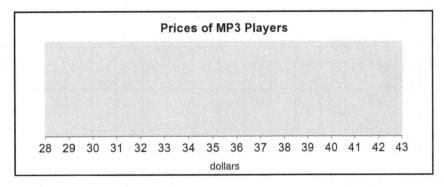

   This distribution is _____-shaped. The mean is _____ and describes the peak of the distribution well/somewhat/poorly (*choose one*).

   The m.a.d. is _____, and it means that on average, the price of an MP3 player differs from the mean price by about _____ dollars.

3. These are the weekly work hours of employees in two places: a restaurant and an office.

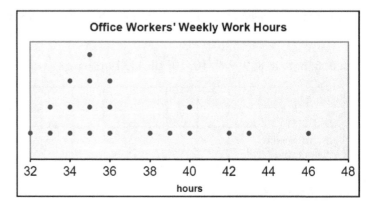

   a. Use the dot plots and determine visually:

   Which group works more hours?

   In which group is there more variability in the weekly work hours?

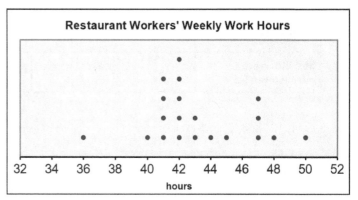

   b. Calculate the range for each group.

   Office workers: _____

   Restaurant workers: _____

   c. Calculate the mean and the m.a.d. for each group. Use the tables below to help.

   Office workers: mean _____   m.a.d. _____

   Restaurant workers: mean _____   m.a.d. _____

   d. Do your calculations support the answers you gave in (a)?

Office workers:

work hours	32	33	33	34	34	34	35	35	35	35	36	36	36	38	39	40	40	42	43	46	*mean*	
absolute difference from mean																					*m.a.d.*	

Restaurant workers:

work hours	36	40	41	41	41	41	42	42	42	42	42	43	43	44	45	47	47	47	48	50	*mean*	
absolute difference from mean																					*m.a.d.*	

Making Histograms

Histograms are like bar graphs, but the bars are drawn so they touch each other. Histograms are used only with numerical data.

Example. These are prices of hair dryers in three stores (in dollars). Make a histogram.

14 15 19 20 20 20 21 24 25 34 34 35 35 37 42 45 55

We need to decide how many bins to make and how "wide" they are. For that, we first calculate the **range**, or the difference between the greatest and smallest data item. It is 55 − 14 = 41. Then we divide the range into equal parts (bins) to get the *approximate* bin width.

If we make five bins, we get 41 ÷ 5 = 8.2 for the bin width. The bins would be 8.2 units apart. However, in this case it is nice to have the bins go by whole numbers, so we round 8.2 up to 9 and use 9 for the bin width.

The important part is that *each data item needs to be in one of the bins*. You may have to try out slightly different bin widths and starting points to see how it works. This time, starting the first bin at 13 makes the last bin to end at 57, which works, because the data will "fit" into the bins. (Starting at 14 would work, too.)

The **frequency** describes *how many data items fall into that bin*. Lastly, all we need is to draw the histogram, remembering that the bars touch each other.

Price ($)	Frequency
13..21	7
22..30	2
31..39	5
40..48	2
49..57	1

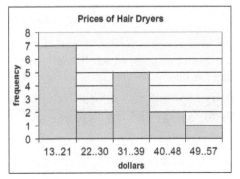

This is a **double-peaked** distribution and **skewed to the right** (the direction of skewness is where the "long tail" of the distribution is, in this case to the right). Since it definitely is *not* bell-shaped, the mean is *not* a good measure of center. Therefore, the *median* is the better choice for measure of center.

Consequently, we need to use *interquartile range*, and not mean absolute deviation, as a measure of variation.

The median is underlined below:

14 15 19 20 20 20 21 24 **25** 34 34 35 35 37 42 45 55

The 1st quartile is $20 and the 3rd quartile is $36 (verify those). So the interquartile range is $16.

This means that half of the data is found between 20 and 36 dollars. This price range is quite large for devices with a median price of only $25. A large range compared to the median describes data that is widely scattered. (We can also see that from the dot plot.)

Since the median is $25, which is nearer the low end of the interval from $20 to $36, the prices are somewhat more concentrated in the lower end of that interval.

For a comparison, look at the dot plot as well. It has a similar shape to the histogram.

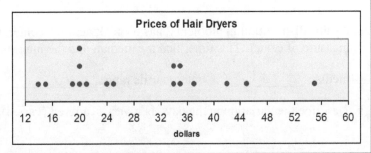

1. This data lists the heights of 24 swimmers in centimeters. Make a histogram with <u>five</u> bins.

   155 155 156 157 158 159 159 160 162 162 163 163
   164 165 166 167 167 168 168 170 172 174 175 177

Height (cm)	Frequency

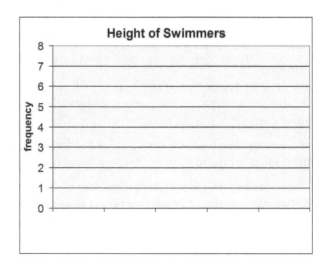

2. Make a histogram from this data, which lists all the scores a basketball team had in the games in one season. Make six bins.

   60 62 68 71 72 72 73 74 74 74 75 75 76 77 77 77
   78 78 78 79 79 81 81 82 83 83 85 86 88 90 92 95

Score	Frequency

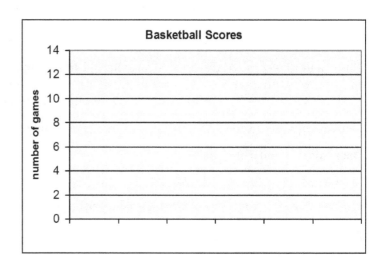

3. **a.** As this distribution in problem 2 has its peak near the center, you could use either mean or median as a measure of center. This time, find the median and the interquartile range.

   median _____   interquartile range _____

   b. Describe the variation in the data. Is it very scattered (varied), somewhat so, or not very much so?

4. This data is the finishing times of a group of athletes for the 100-meter dash. Make two histograms, one with four bins and the other with five, and compare them.

Notice the differences in how they look. Both show a peak, but the one with more bins shows more detail. The look of a histogram changes depending on the number of bins and the starting point of the first bin. For this reason, if the data set is small, a dot plot or a stem-and-leaf plot may work better than a histogram.

a. Make a histogram with 4 bins.

11.8 12.0 12.1 12.1 12.3 12.4 12.5 12.5 12.6 12.6 12.7
12.7 12.7 12.7 12.7 12.8 12.8 12.8 12.8 12.9 12.9 13.1

Finishing times (seconds)	Frequency

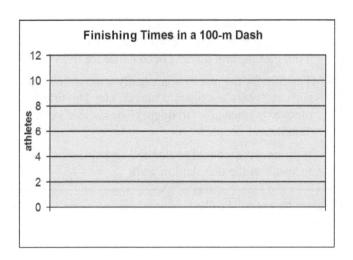

b. Make a histogram with 5 bins (from the same data).

11.8 12.0 12.1 12.1 12.3 12.4 12.5 12.5 12.6 12.6 12.7
12.7 12.7 12.7 12.7 12.8 12.8 12.8 12.8 12.9 12.9 13.1

Finishing times (seconds)	Frequency

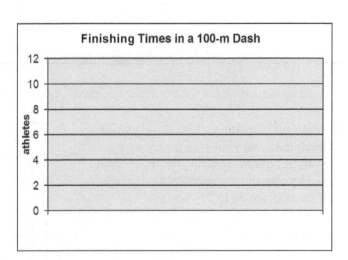

Puzzle Corner Find a quick, *mental math* method for calculating the mean for this data set. 102, 94, 99, 105, 96, 107, 101, 104 (the weights of a litter of kittens at birth, in grams)

Summarizing Statistical Distributions

Here are some basics things that you can do to summarize the information in a data set:

- Explain in a sentence or two what was measured and how it was measured.
- Indicate how many observations (data items) there are.
- Create a graph.
- Give a title to your graph.
- Give titles to the axes in your graph, unless it is very clear what the graph is about even without them. For example, a dot plot usually doesn't need a title for the vertical axis.

  The title for the one axis is usually a measurement unit, such as "years", "dollars", "days", etc. The title for the other axis is usually "frequency" or similar, such as "number of people".

- Indicate a measure of center and its value that describes the peak of the distribution well.

If the data is numerical, also:

- Describe the overall shape or pattern of the distribution, including any striking deviations from the overall pattern (outliers, gaps, additional minor peaks/clusters).
- Choose a measure of variation based on the chosen measure of center, and indicate its value.

Here's an example of how your summary might look like. Note that this particular example does not have gaps, outliers, or clusters.

Summary: The parents of a total of 36 elementary school children were asked how many hours their child had slept the previous night. Most of the children slept 9 or 10 hours, the median being 9 hours. The distribution is asymmetrical and left-skewed as there were some children who slept as little as 6 hours. The interquartile range is from 8 to 10 hours, or 2 hours.

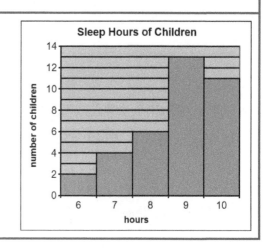

You may use a calculator and a spreadsheet program (if available) in all the questions in this lesson.

1. Make up a situation where this data was measured, and write a summary to go with this graph. To help you, the median is 38 years and the interquartile range is 3 years (from 36 to 39 years).

   Summary:

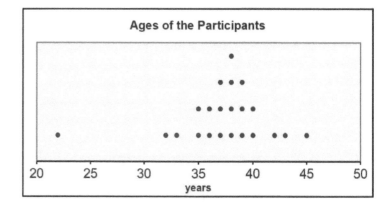

2. These are the lengths of words on three pages in a certain children's storybook. The data is already sorted:

   1 1 2 2 2 2 2 2 2 2 3 3 3 3 3 3 3 3 3 3 3 3 3 3 4 4 4
   4 4 4 4 4 4 4 4 4 5 5 5 5 5 5 5 5 5 5 6 6 6 6 7 7 8 8 8

Make a graph of your choosing, and give a summary of this data.

Summary:

3. Make a <u>histogram</u> from this data, and give a summary.

Life expectancy at birth (years)	Frequency
65.0-67.9	1
68.0-70.9	0
71.0-73.9	5
74.0-76.9	8
77.0-79.9	6
80.0-82.9	3

Country	Life expectancy at birth (years)
Canada	82.2
Costa Rica	80.8
Chile	80.7
Peru	79.9
Colombia	79.3
Panama	79.3
United States	78.5
Ecuador	78.4
Uruguay	77.1
Argentina	76.6
Trinidad and Tobago	76.1
Mexico	76
Brazil	75.9
Paraguay	75.8
El Salvador	75
Nicaragua	75
Belize	74.4
Venezuela	73.9
Bolivia	72.1
Guatemala	72
Honduras	71.9
Suriname	71.8
Guyana	65.7

(source: WHO; 2019 data)

Summary:

4. (An optional challenge) The table below lists the world's 25 tallest buildings and their heights. Make a graph of your choosing, and give a summary of this data. Use grid paper or a spreadsheet program. Before you start, think which type of graph—a dot plot, a boxplot, or a histogram—would best portray this data.

The tallest (completed and topped out) buildings in the world				
Rank	Building	City	Height (m)	Built
1	Burj Khalifa	Dubai	828 m	2010
2	Shanghai Tower	Shanghai	632 m	2014
3	Makkah Royal Clock Tower Hotel	Mecca	601 m	2012
4	One World Trade Center	New York City	541.3 m	2013
5	Taipei 101	Taipei	509 m	2004
6	Shanghai World Financial Center	Shanghai	492 m	2008
7	International Commerce Centre	Hong Kong	484 m	2010
8	Petronas Tower 1	Kuala Lumpur	452 m	1998
8	Petronas Tower 2	Kuala Lumpur	452 m	1998
10	Zifeng Tower	Nanjing	450 m	2010
11	Willis Tower (Formerly Sears Tower)	Chicago	442 m	1973
12	Kingkey 100	Shenzhen	442 m	2011
13	Guangzhou International Finance Center	Guangzhou	440 m	2010
14	Marina 101	Dubai	432 m	2014
15	Trump International Hotel and Tower	Chicago	423 m	2009
16	Jin Mao Tower	Shanghai	421 m	1999
17	Princess Tower	Dubai	414 m	2012
18	Al Hamra Firdous Tower	Kuwait City	413 m	2011
19	2 International Finance Centre	Hong Kong	412 m	2003
20	23 Marina	Dubai	395 m	2012
21	CITIC Plaza	Guangzhou	391 m	1997
22	Shun Hing Square	Shenzhen	384 m	1996
23	Central Market Project	Abu Dhabi	381 m	2012
24	Empire State Building	New York City	381 m	1931
25	Elite Residence	Dubai	380.5 m	2012

Source: Wikipedia

Stem-and-Leaf Plots

This lesson is optional.

A stem-and-leaf plot is made using the numbers in the data, and it looks a little bit like a histogram or a dot plot turned sideways.

In this plot, the tens digits of the individual numbers become the **stems**, and the ones digits become the **leaves**. For example, the second row 2 | 1 2 5 8 actually means 21, 22, 25, and 28. Notice how the leaves are listed in order from the smallest to the greatest.

Ages of the participants in the County Fair Karaoke Contest:

14 18 21 22 25 28 30 30
31 33 33 36 37 40 45 58

Stem	Leaf
1	4 8
2	1 2 5 8
3	0 0 1 3 3 6 7
4	0 5
5	8

4 | 5 means 45

Since stem-and-leaf plots show not only the *shape* of the distribution but also the individual *values*, they can be used to get a quick overview of the data. This distribution has a central peak and is somewhat skewed to the right.

You can also find the median fairly easily because you can follow the individual values from the smallest to the largest, and find the middle one.

Stem-and-leaf plots are most useful for numerical data sets that have 15 to 100 individual data items.

1. **a.** Complete the stem-and-leaf plot for this data:

    19 20 34 25 21 34 14 20 37 35 20 24 35 15 45 42 55

    (prices of hair dryers in three stores)

    b. What is the median?

Stem	Leaf
1	
2	
3	
4	
5	

    5 | 4 means 54

2. **a.** Complete the stem-and-leaf plot for this data. This time, the stems are the first two digits of the numbers, and the leaves are the last digits.

    709 700 725 719 750 740 757 745 786 770 728 755

    (monthly rent, in dollars, for one-bedroom apartments in Houston, Texas)

    b. Find the median monthly rent.

    c. Find the interquartile range.

    d. Describe the spread of the distribution
    (is the data spread out a lot, a medium amount, a little, etc.)

Stem	Leaf
70	
71	
72	
73	
74	
75	
76	
77	
78	
79	

    71 | 9 means 719

3. The data below gives the average days with precipitation for cities in Texas. For example, the number 69 for Amarillo means that generally, the city of Amarillo gets at least some rain on 69 days out of 365 in a year.

   a. Make a stem-and-leaf plot of the data.

City	Days with rain (average)	City	Days with rain (average)
Abilene	67	Houston	106
Amarillo	69	Lubbock	63
Austin	84	Midland-Odessa	52
Brownsville	73	Pt Arthur Beaumont	105
Corpus Christi	77	San Angelo	59
Dallas/Ft. Worth	79	San Antonio	82
Del Rio	63	Victoria	90
El Paso	49	Waco	79
Galveston	96	Wichita Falls	71

Stem	Leaf

   8 | 4 means 84

   b. Find the median and the interquartile range.

4. This stem-and-leaf plot shows the scores given to a gymnast in a contest by 10 different judges. Now the leaves are the first decimal digits of the numbers.

   a. What is the median of the scores?

   b. What is the mean?

   c. Which of the two measures of center describes the center of this distribution better?

Stem	Leaf
9	1 2
9	3 3
9	5 5 5 6 6
9	7

   9 | 7 means 9.7

5. a. Make a stem-and-leaf plot of this data.

   254 248 232 255 250 227 235 238 260 231 259

   (Results of a children's long-jump contest, in centimeters)

   b. Describe the shape of the distribution.

Stem	Leaf

   c. Which measure of center (mean, median, or mode) would best describe this data? Why?
   (You do not have to calculate them, though you may.)

   25 | 9 means 259

Chapter 10 Mixed Review

1. Find the greatest common factor of the given number pairs. (The Greatest Common Factor/Ch.6)

   a. 87 and 36　　　　　　　　　　　　　　**b.** 96 and 16

2. Find the least common multiple of the given number pairs. (The Least Common Multiple/Ch.6)

   a. 6 and 12　　　　　　　　　　　　　　**b.** 8 and 12

3. First, find the GCF of the numbers. Then factor the expressions using the GCF. (Factoring Sums/Ch.6)

a. The GCF of 72 and 12 is _____ 12 + 72 = ___ (___ + ___)	**b.** The GCF of 42 and 66 is _____ 42 + 66 = ___ (___ + ___)

4. **a.** The points (−8, 7), (−5, 3) and (4, 0) are vertices of a triangle. Draw the triangle.
   (Coordinate Grid/Ch.8)

   b. Move the triangle five units down *and* three units to the right. Notice there are *two* movements! Write the coordinates of the moved vertices.

   (−8, 7) → (_____ , _____)

   (−5, 3) → (_____ , _____)

   (4, 0) → (_____ , _____)

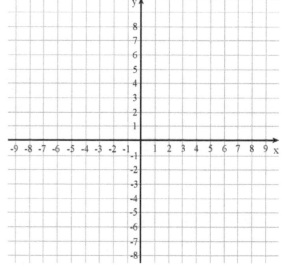

5. Write an equation for each situation—even though you could easily solve the problem without an equation! Lastly, *solve* the equation you wrote. (Writing Equations/Ch.2)

a. The area of a rectangle is 304 m² and one of its sides is 19 m. How long is the other side?

b. Mike weighed five identical books on the scales. They weighed 6.7 kg. What was the weight of one book?

6. Simplify the expressions. (Writing and Simplifying Expressions 1: Length and Perimeter/Ch.2)

a. $z \cdot z \cdot z \cdot 7$	b. $8 \cdot a \cdot 3 \cdot b \cdot 10$
c. $2 + x + x + x + x$	d. $5t - 2t + 6$

7. Add or subtract the fractions. Give your answer as a mixed number. (Revision: Add and Subtract Fractions/Ch.7)

a. $\dfrac{5}{11} + \dfrac{1}{2} + \dfrac{5}{6}$

b. $3\dfrac{11}{12} - \dfrac{5}{10} + \dfrac{1}{4}$

8. What part of a whole pizza is two-thirds of nine-tenths of a pizza?
   (Revision: Multiplying Fractions 1/Ch.7)

9. A piglet is born weighing 3 lb 4 oz. If it gains approximately
   7 1/3 ounces per day during its 12-day nursing period, then
   how much will it weigh at weaning (the end of the nursing period)?
   (Convert Customary Measuring Units/Ch.3)

10. There are 1 3/5 chocolate bars left, and you want to share
    that amount equally between you and your three friends.
    How much will each of you get?
    (Divide Fractions/Ch.7)

11. Simplify. In (e), write using a number. (Integers/Ch.8)

    a. $|9|$ b. $|-3|$ c. $|0|$ d. $-(-28)$ e. the opposite of -7

12. Write an addition or subtraction sentence. (Addition and Subtraction as Movements/Ch.8)

    a. You are at ⁻12. You jump 7 steps to the right. You end up at _____.

    b. You are at 2. You jump 8 steps to the left. You end up at _____.

13. On a separate sheet of paper, draw a right triangle with an *area* of 8 square inches.
    (Area of Right Triangles/Ch.9)

14. Find the total area of the boat and its sail.
    (Polygons in the Coordinate Grid/Ch.9)

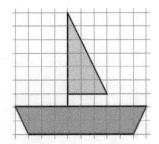

15. Find the area of the yellow shaded figure at the right.
    (Area of Polygons/Ch.9)

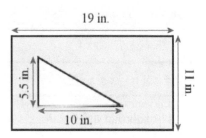

16. This graph shows the personal income of people in the United States, 15 years and older, in 2005. The graph does not show the approximately 13,000 people who earned more than $100,000. (The "k" in the chart means $1,000. So, 50k means $50,000.)
    (Understanding Distributions/Ch.10; What Percentage...?/Ch.5)

    a. Fill in. This is _____ - shaped distribution which means it is very asymmetrical.

    b. Estimate from the graph about how many people belong to the first column on the graph (in other words, they earned less than $12,500 in 2005).

    c. Estimate from the graph about how many people earned between $12,500 and $25,000.

    d. Now use your answers from (b) and (c), estimation, and the fact that the total number of people is 198,617,000. *Approximately* what percentage of people earned less than $25,000?

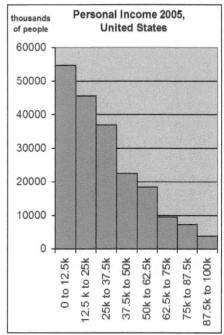

Source: Census.gov

Puzzle Corner Be a teacher-detective: How did the children come up with these answers?

a. Jerry cannot figure out what went wrong:	**b.** Emily has something strange going on here:
$\frac{2}{7} \div 1\frac{3}{4} = 6\frac{1}{8}$	$\frac{4}{5} \div 1\frac{1}{2} = 1\frac{3}{5}$
$2\frac{1}{3} \div \frac{2}{5} = \frac{6}{35}$	$2\frac{1}{3} \div \frac{1}{4} = 1\frac{1}{3}$
$1\frac{1}{5} \div 2\frac{2}{3} = 2\frac{2}{9}$	$1\frac{1}{5} \div 2\frac{2}{3} = \frac{3}{10}$
What error did Jerry make each time?	What error did Emily make each time?

Statistics Review

You may use a calculator in all questions in this lesson.

1. Are these statistical questions or not? If not, change the question so that it becomes a statistical question.

   a. Which kind of books do the visitors of this library like the best?

   b. How many pages are in the book *How to Solve It* by G. Polya?

2. **a.** Find the mean, median and mode.
   Hint: recreate the list of the original data.

   Mean:

   Median:

   Mode:

   b. Describe the shape and any striking features of the distribution.

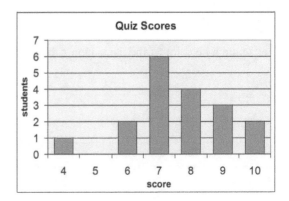

3. This graph shows the hourly wages in euros per hour of the 89 employees in the Inkypress Print Shop.

   a. *About* what fraction of the people earn 7-8 euros/hour?

   b. Describe the shape and any striking features of the distribution.

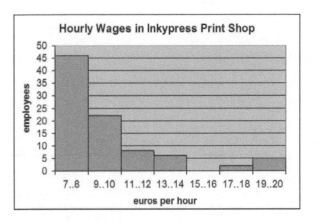

   c. The mean is 9.66 euros/hour and the median is 8 euros/hour. Which is better in describing the majority's wages in this print shop?

   d. Which measure of variation should be used to describe this data, based on your answer to (c)?

179

4. **a.** Find the five-number summary and the interquartile range of this data set.

   2, 5, 5, 6, 6, 7, 7, 7, 8, 8, 8, 9, 12

   Minimum _____ First quartile _____ Median _____ Third quartile _____ Maximum _____

   Interquartile range: _____

   b. Make up a situation where this data could have come from.

   c. Make a boxplot of the data. Don't forget to give it a title.

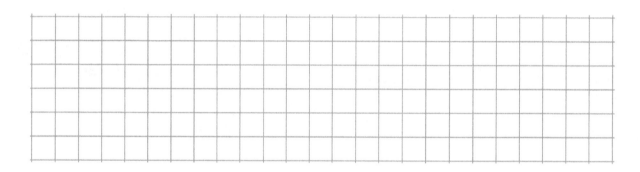

5. **a.** Make a dot plot for this set of data: 0, 1, 2, 2, 3, 3, 4, 4, 4, 4, 5, 5, 5, 6, 7, 8, 10, 13, 14, 20
   (number of sick days the employees of a small company had last year).

   b. Calculate the mean and the mean absolute deviation. You can use the table below to help.

														mean	
abs. difference from mean														*m.a.d.*	

6. a. This data gives you the average number of days with precipitation in a year in 32 major European cities. For example, on average, in Amsterdam it rains (at least a little bit) 132 days out of every 365 days.

Create a histogram from this data. Make six bins.

Days with Rain	Frequency

City	Days with Rain
Athens, Greece	43
Madrid, Spain	63
Bucharest, Romania	72
Lisbon, Portugal	77
Rome, Italy	78
Budapest, Hungary	81
Sofia, Bulgaria	81
Istanbul, Turkey	84
Warsaw, Poland	93
Prague, Czech Republic	94
Zagreb, Croatia	95
Tirana, Albania	98
Vienna, Austria	98
Kiev, Ukraine	99
Copenhagen, Denmark	102
Stockholm, Sweden	105
Berlin, Germany	106
London, UK	109
Paris, France	111
Oslo, Norway	113
Helsinki, Finland	115
Riga, Latvia	120
Moscow, Russia	121
Luxembourg, Luxembourg	122
Vilnius, Lithuania	122
Zurich, Switzerland	125
Tallinn, Estonia	127
Dublin, Ireland	129
Hamburg, Germany	129
Amsterdam, Netherlands	132
Reykjavík, Iceland	148
Brussels, Belgium	199

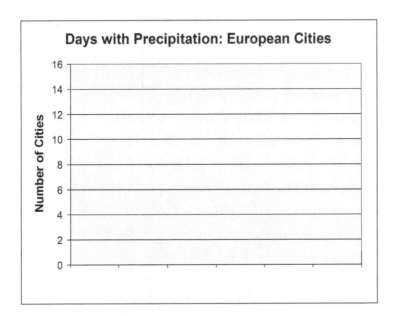

b. Describe the shape and any striking features of the distribution.

c. Choose a measure of center that describes the data well, and determine its value.

d. (Optional.) This data can also easily be plotted in a stem-and-leaf plot. Make a stem-and-leaf plot and compare it visually to the histogram you made.

7. Make up a situation where this data was measured, and write a summary to go with this graph.

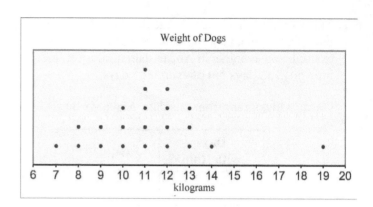

8. **a.** Make a stem-and-leaf plot of this data.

   78 82 84 75 90 66 77 64 112 84 85 92

   (The height of a group of toddlers, in centimeters.)

   b. Find the median.

   c. Describe the shape and any striking features of the distribution.

Stem	Leaf

8|2 = 82